Virtual Worlds

New England Complex Systems Institute Series on Complexity

YANEER BAR-YAM, EDITOR-IN-CHIEF

Unifying Themes in Complex Systems, Volume I
Edited by Yaneer Bar-Yam

Unifying Themes in Complex Systems, Volume II
Edited by Yaneer Bar-Yam

Virtual Worlds
Edited by Jean-Claude Heudin

Virtual Worlds

Synthetic Universes, Digital Life, and Complexity

Jean-Claude Heudin, Editor

Westview
PRESS
Advanced Book Program
A Member of the Perseus Books Group

Hardcover edition first published in 1999 in the United States of America by Perseus books; paperback edition published in 2004 by Westview Press, a Member of the Perseus Books Group.

Find us on the world wide web at www.westviewpress.com

Westview Press books are available at special discounts for bulk purchases in the United States by corporations, institutions, and other organizations. For more information, please contact the Special Markets Department at the Perseus Books Group, 11 Cambridge Center, Cambridge, MA 02142, or call (800) 255-1514 or (617) 252-5298, or e-mail special.markets@perseusbooks.com.

Library of Congress Catalogue Card Number: 98-89113
ISBN 0-8133-4286-4 (paperback)

The paper used in this publication meets the requirements of the American National Standard for Permanence of Paper for Printed Library Materials Z39.48–1984.

10 9 8 7 6 5 4 3 2 1

Contents

Foreword

The story of this book began in June 1997. As a researcher interested in the sciences of complexity, I was convinced that a promising approach to study complexity was to simulate the evolution of "worlds" considered as wholes, that is with their own "physical" and "biological" laws. In the last few years, there has been an increasing interest in the design of virtual environments using advanced three-dimensional image synthesis and dedicated interfaces such as head mounted displays and data gloves. At the same time, the field of Artificial Life has addressed fundamental questions about the nature of the living by synthesizing life-like phenomena on computers. However, only few works have used together these two approaches in order to create realistic virtual worlds. Of course, such multidisciplinary projects seem extremely difficult to do in any sort of complete and realistic manner, but I think they can be developed at many scales and benefit from a large amount of previous works in various fields. In this framework, I decided to organize the First International Conference on Virtual Worlds in July 1998. It brought together more than 150 scientists and artists involved in Virtual Reality, Artificial Life, Multi-agent Systems and other disciplines of Computer Sciences and Electronic Art, who shared a common interest in the synthesis of virtual worlds on computers. The proceedings of the conference were managed in a classical way and published by Springer, including all the papers that were presented at the conference. The goal was to capture in print the stimulating mix of ideas and works that were presented.

The goal of this volume is quite different, and is complementary with the conference proceedings. While the proceedings represented a "screenshot" of this new emerging trend of research, including a wide spectrum of works, this book is more centered on my initial ideas when I launched the conference. In this perspective, the selected contributions to this volume have been heavily reviewed in order to reflect these ideas. Also, this highly selective filtering process has been necessary to establish a standard for the highest quality of published research required by such a new emerging field. The resulting volume is composed of ten chapters, each written by different contributors working on a specific aspect of Virtual Worlds.

I wrote the first chapter and simply called it "Virtual Worlds". This chapter introduces Virtual Worlds as a new field of research in the framework of the Sciences of Complexity. After a short historical background, it establishes its foundational principles and gives two examples of Virtual Worlds experiments. Then, it introduces some related philosophical issues and its relationships with Cyberart.

The second chapter, named "An Evolutionary Approach to Synthetic Biology: Zen and the Art of Creating Life", was written by Tom Ray in 1994 and was originally published in the first volume of the Artificial Life journal. I wanted to include a chapter written by Tom since I consider him as a pioneer for this new emerging field and his work on Tierra has greatly influenced a large number of researchers.

The third chapter, called "Animated Artificial Life", addresses important questions raised by the animation and evolution of digital creatures. Written by Jeffrey Ventrella, it describes a collection of fascinating experiments including "Darwin Ponds", a two-dimensional virtual world in which "sexual swimmers" live and evolve.

The fourth chapter is written by Nadia Magnenat-Thalmann and Laurent Moccozet. "Virtual Humans on Stage" discusses the very difficult and challenging problem of creating virtual humans with the highest possible realism in rendering and animation. Nadia is recognized as the leading expert in this trend of research worldwide.

Chapter five is written by Bruce Damer, Stuart Gold, Karen Marcelo and Frank Revi of the Contact Consortium. "Inhabited Virtual Worlds in Cyberspace" addresses another important issue of Virtual Worlds: its relationships with Internet and the World-Wide-Web. It seems clear that Cyberspace represents the ideal medium for synthesizing huge virtual worlds accessible from a large community of users. Bruce and his colleagues are doing a wonderful job in the cyberspace as well as in the real world for promoting Virtual Worlds experiments and applications.

The sixth chapter describes one of the most challenging virtual world projects using state-of-the-art Virtual Reality and Artificial Life technologies. The "Virtual Great Barrier Reef" was first presented at the Virtual Worlds conference in Paris by Scot Refsland. I must point out that some of the contributions to the actual project were the result of discussions and exchanges that occurred during the conference.

The seventh chapter is written by Yaneer Bar-Yam, president of the New England Complex System Institute. His chapter, named "Virtual Worlds and Complex Systems", exemplifies the strong relationships between Virtual Worlds and the sciences of complexity. Yaneer was one of the first researchers to support my initiative when I launched the Virtual Worlds conference.

Chapter eight illustrates a virtual world project designed for studying complexity. "Investigating the Complex with Virtual Soccer", written by Itsuki Noda and Ian Frank, describes Virtual Soccer as a framework well adapted for studying complex phenomena and interactions between a set of interacting agents. Virtual Soccer could play for Virtual Worlds the same role as chess played for Artificial Intelligence.

The ninth chapter is written by Rodney Berry, an australian artist working on Virtual Worlds. "Feeping Creatures" is a fascinating virtual world mixing VRML-based Virtual Reality, Artificial Life and three-dimensional sound generation. I asked him to write a chapter focussing on sound since I am convinced that sound and music will play a crucial role in the design of realistic and imaginary virtual worlds.

The last chapter, named "Art in Virtual Worlds: Cyberart", is written by Olga Kisseleva who greatly helped me in organizing the conference. It is not a classical scientific chapter, but rather the look of an artist over Virtual Worlds considered as a major new medium based on the collaboration of science, technology and art.

It is an exciting time to be involved in the synthesis of a new virtual universe. I would like to thank all the contributors to this volume and every one involved so far in the great adventure of Virtual Worlds. My sincere hope is that this book will contribute by helping to establish this new emerging trend of research as a legitimate field of scientific inquiry.

Jean-Claude Heudin, Paris, October 1998.

"In another moment Alice was through the glass, and had jumped lightly down into the looking-glass room. The very first thing she did was to look whether there was a fire in the fireplace, and she was quite pleased to find that there was a real one, blazing away as brightly as the one she had left behind. [...] Then, she began looking about, and noticed that what could be seen from the old room was quite common and uninteresting, but that all the rest was as different as possible. For instance, the pictures on the wall next the fire seemed to be all alive [...]."

Through the looking glass – and what Alice found there
Lewis Carroll, 1872

Chapter 1

Virtual Worlds

Jean-Claude Heudin
International Institute of Multimedia
Pôle Universitaire Léonard de Vinci
Jean-Claude.Heudin@devinci.fr
http://www.devinci.fr/iim/jch

1.1. Introduction

Imagine a virtual world with digital creatures that looks like real life, sounds like real like and even feels like real life. Imagine a virtual world with not only nice three-dimensional graphics and animations, but also with realistic "physical" laws and forces. This virtual world could be familiar, reproducing some parts of our reality, or unfamiliar, with strange "physical" laws and artificial life forms.

In this chapter, we introduce Virtual Worlds as a new field of research that studies complexity by attempting to synthesize digital universes on computers. This includes models of simple abstract worlds such as cellular automata to more sophisticated virtual environments using Virtual Reality and Artificial Life techniques. First, we make a brief historical review of these two fields and we suggest that experiments taking advantages of both approaches represent a new and promising trend of research. Then, we describe two examples of virtual worlds. The first example is one of the simplest virtual universe that we can imagine based on a two-dimensional two-state cellular automaton. The second example shows a more sophisticated world using an immersive three-dimensional environment where digital creatures "live" and evolve. Finally, we discuss some of the philosophical implications of Virtual Worlds. We conclude by outlining some of the possible outcomes of this new field of scientific inquiry.

1.2. Virtual Worlds

1.2.1. The Synthesis of "Real" and Imaginary Universes

In the last few years, there has been an increasing interest in the design of artificial environments using image synthesis and Virtual Reality (VR). The emergence of industry standards such as VRML [Hartman 1996] is an illustration of this growing interest. During the same period of time, the field of Artificial Life (AL) has addressed the study of complex phenomena such as self-organization, reproduction, development and evolution of artificial life-like systems. However, very few works have used an AL approach together with advanced three-dimensional (3D) graphics or VR techniques. Considering recent advances in both fields, catalyzed by the development of Internet, a unified approach seems to be one of the most promising trend of research for the synthesis of realistic and imaginary virtual worlds.

1.2.2. Virtual Reality

The roots of VR may be traced back to the early 1940s, when an entrepreneur by the name of Edwin Link developed flight simulators for forces in order to reduce training times and costs. The early simulators were complex mechanical systems with a relatively poor illusion of flight. In 1965, Ivan Sutherland published a paper called "The Ultimate Display" in which he described the computer as "a window through which one beholds a virtual world" [Sutherland 1965]. The term "Artificial Reality" was first coined by Myron W. Krueger in the mid 1970s to cover the Videoplace project [Krueger 1991] and the head-mounted 3D viewing technology that originated with Ivan Sutherland. During the following years, the terms "Virtual Cockpit", "Virtual Environment" and "Virtual World" were used to describe specific projects. In 1989, Jaron Lanier, CEO of Virtual Presence Ltd., coined the term "Virtual Reality" to bring all these "virtual" projects under a single field of research. It then refers typically to 3D graphical environments with stereo viewing goggles and reality gloves. In 1994, the Virtual Reality Modeling Language (VRML) was just a concept. Presently, it is a *de facto* standard that allows to describe objects and combine them to create interactive simulations that incorporate 3D real-time graphics, motion physics and multi-user participation on the World-Wide-Web [Hartman 1996].

A first approach in categorizing Virtual reality experiments leads to consider four basic classes [Verna 1998]:

1. *Virtual Reality* refers to the modeling of an existing real environment and visualizing it in 3D.

2. *Augmented Reality* means adding virtual information or objects which not belong to the original scene, like virtual objects included in real-time onto a live video.

3. *Released Reality* releases real world constraints, like for instance the inability to reverse time or to escape from the law of gravitation.

4. *Artificial Reality* refers to as the ability to design worlds that do not exist and to display them in 3D.

In all of these approaches, besides displaying 3D spaces, two other important concepts are involved: *immersion* and *interaction*. In state-of-the-art VR, the operator evolves in the generated world thanks to data suit, head mounted display and data gloves. These special input/output devices create the experience of being immersed in the virtual environment.The idea is to feel "physically" present in the virtual environment and to interact with it. In order to keep a coherent feeling of presence, an ideal immersive system should "disconnect" the operator from its real environment.

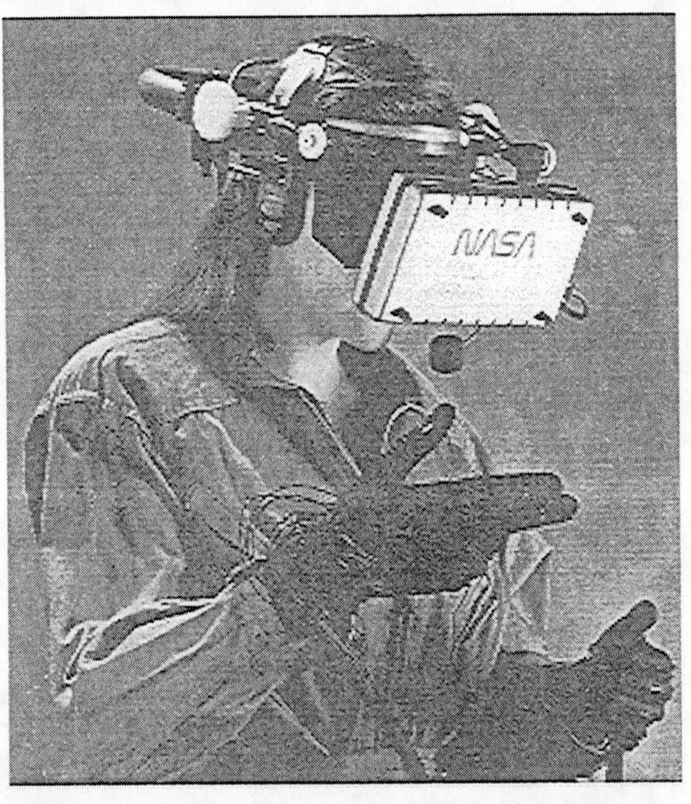

Figure 1. The VIEW system developed at the NASA Ames Research Center under the direction of Scott Fisher.

1.2.3. Artificial Life

Artificial Life is a new field of research initiated by Christopher Langton at a workshop held at Los Alamos in 1987 [Langton 1988]. He coined Artificial Life as "the study of man-made systems that exhibit behaviors characteristic of natural living systems, such as self-organization, reproduction, development and even evolution. It complements the traditional biological sciences concerned with the analysis of living organisms by attempting to synthesize and study life-like behaviors within computers or other "alternative" media. By extending the empirical foundation upon which biology rests beyond the carbon-chain-based life that has evolved on Earth, Artificial Life can contribute to the theoretical biology by locating "life-as-we-know-it" within the larger context of "life-as-it-could-be", in any of its possible physical incarnations" [Langton 1994].

It exists a large diversity of trends that have been taken and AL seems to be best described by a list of related research programs: Cellular Automata, Biomorphs and

L-Systems, Autocatalytic Networks, Simulation of Ecological Systems, Bio-inspired Multi-agent Systems, Evolutionary Computing, Evolving Robots, Evolvable Hardware, Artificial Nucleotides, and many others including related philosophical issues. As one could understand, there is a large number of possible approaches concerned with attempts to synthesize life-like phenomena. However, the key concept in all these works is *emergent behavior*. Thus, the general approach is rather bottom-up modeling than working analytically downward from a complex all to a set of simpler components. AL starts at the bottom, viewing a system as a large population of simple agents, and works upwards synthetically, constructing large aggregates of autonomous rule-governed agents which interact with one another nonlinearly.

In this framework, we can sum up the essential features of an AL model by the following points [Langton 1988]:

1. The model consists of a population of simple agents.
2. There is no single agent that directs all of the other agents.
3. Each agent details the way in which a simple entity reacts to local situations in its environment, including encounters with other entities.
4. There is no rule in the system that dictate global behaviors.
5. Any behavior at levels higher than the individual agents is therefore emergent.

As should be expected, in recent years, there has been two shifts in emphasis. In the first shift, AL studies are characterized by more connections to real systems exemplified by the growing number of works in Evolutionary Robotics [Brooks 1994] and Evolvable Hardware [Sanchez 1996]. In the second shift, researchers design more sophisticated artificial worlds where evolving population are studied, including models of physical dynamics. This second shift of AL along with VR represent the two roots of the emerging Virtual Worlds approach.

1.2.4. Early Virtual Worlds

In a sense, the first approach to the synthesis of a virtual world was due to the mathematician John Conway. Following the works done by John von Neumann on Cellular Automata and self-reproduction [Von Neumann 1966], he designed a computer-based cellular model of a simple two-dimensional world called "the Game of Life" [Gardner 1970] which has open the road to a large number of experiments (cf. section 2).

More recently, Tom Ray has designed Tierra, an environment producing synthetic organisms based on a computer metaphor of organic life in which CPU time is the "energy" resource and memory is the "material" resource [Ray 1991]. Memory is organized into informational patterns that exploit CPU time for self-replication. Mutation generates new forms, and evolution proceeds by natural selection as different creatures compete for CPU time and memory space (cf. chapter 2).

Karl Sims was probably one of the first researcher using a similarly flexible genetic system to specify solutions to the problem of being a "creature" built of collections of blocks, linked by flexible joints powered by "muscles" controlled by circuits based on an evolvable network of functions [Sims 1995]. Sims embedded these "block creatures" in simulations of real physics, such as in water or on a surface.

These experiments produced a bewildering and fascinating array of creatures, like the swimming "snake" or the walking "crab".

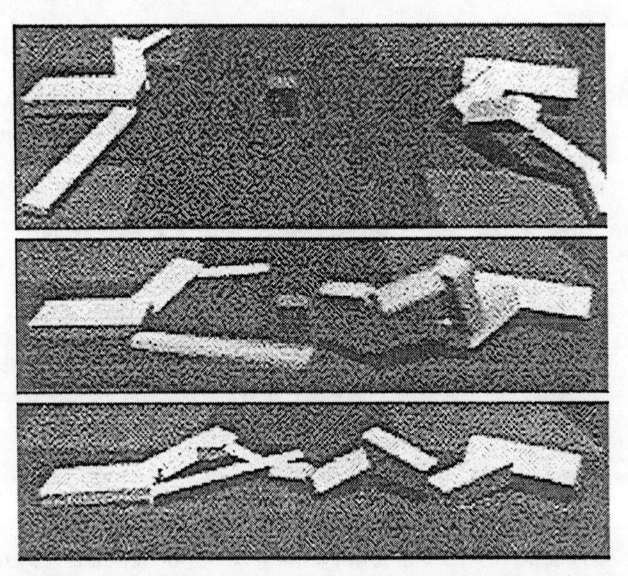

Figure 2. An example of Karl Sims's evolved competing creature. These images show three phases of a competition to gain control of a single cube placed in the center of the world.

Dimitri Terzopoulos and his colleagues have created a virtual marine world inhabited by realistic artificial fishes [Terzopoulos 1995]. They have emulated not only the appearance, movement and behavior of individual animals, but also the complex group behaviors evident in many aquatic ecosystems. Each animal was modeled holistically as an autonomous agent situated in a simulated physical world.

Besides these well-known works, there has been a large number of experiments based on the idea of creating a virtual world on a computer : PolyWorld [Yaeger 1993], Avida [Adami 1994], Darwin Pond [Ventrella 1998] (cf. Chapter 3) and many others.

1.2.5. Virtual Worlds Experiments

In the last few years, the term "Virtual Worlds" has mainly referred to VR applications or experiences. We extend here the use of this term to describe experiments that deal with the general idea of synthesizing digital worlds on computers [Heudin 1998b]. Thus, Virtual Worlds (VW) could be defined as the study of computer programs that implement digital worlds as wholes with their own "physical" and "biological" laws. Constructing such complex artificial worlds seems to be extremely difficult to do it in any sort of complete and realistic manner. Such a new discipline must benefits from a large number of works in various fields: advanced 3D Graphics and Virtual Reality, Artificial Life, Evolutionary Computation, Simulation of Physical Systems, and more. Whereas VR has largely concerned itself with the design of 3D graphical spaces and AL with the simulation of living organisms, VW is concerned with the simulation of worlds considered as wholes and the synthesis of digital universes.

Besides its applications in many fields such as computer games or simulation, this approach is something broader and more fundamental, and can contribute to a better understanding of our real universe. Throughout the natural world, at any scale, from particles to galaxies, one can observe phenomena of great complexity. Research done in traditional sciences such as physics and biology has shown that the basic components of complex systems are quite simple. It is now a crucial problem to elucidate the universal principles by which large numbers of simple components, acting together, can self-organize and produce the complexity observed in our universe. In this framework, VW is also concerned with the formal basis of synthetic universes. Thus we can consider that the synthesis of virtual worlds offers a new approach for studying complexity.

1.3. Example of a Simple Two-dimensional Virtual World

1.3.1. The Life Set

The two-dimensional cellular automaton (CA) Life, originally discovered by John Conway, has been presented in a large number of publications [Gardner 1970]. In this system, the world is a two-dimensional lattice with a finite-state automata at each lattice site. Each automaton, or cell, takes as input the states of its eight neighbors. It will then turn "on" if exactly three of its neighbors are "on" and it will stay "on" so long as either two or three of its neighbors are "on", otherwise it will turn "off". Life is one of the simplest world that one could imagine, but it shows complex dynamics [Poundstone 1985] and has been proven capable of supporting universal computation [Berlekamp 1982]. According to several authors [Packard 1985][McIntosh 1990], Life seems to be an exception in the huge space of possible two-dimensional transition rules. Few variants have been mentioned [Berlekamp 1982][Packard 1985] but never clearly identified or carefully described. Therefore, most researchers have concluded that Life is in fact unique.

The most recent variation on this theme has been a series of articles by Carter Bays [Bays 1987, 1988, 1990, 1992]. Bays has expanded Life into three dimensions and proposed a general definition for the set of possible rules. Each rule can be written in the form $E_bE_hF_bF_h$ where E_b is the minimum number of "living" neighbor cells that must touch a currently living cell in order to guarantee that it will remain alive in the next generation. F_b is the minimum number touching a currently dead cell in order that it will come to life in the next generation and E_h and F_h are the corresponding upper limits. These rules are called the "environment" and "fertility" rules. According to this notation, Conway's Life would be written "Life 2333", that is $E_b = 2$, $E_h = 3$, $F_b = 3$, and $F_h = 3$. This definition can be rewritten in a more formal way by the following (equivalent) relations. Let S_t be the state of an arbitrary cell at generation t, and N_t the number of its living neighbors. Then, the general Life transition rule can be defined by:

$$\begin{array}{lll} \text{if} & (N_t \geq F_b) \,\&\, (N_t \leq F_h) \rightarrow S_{t+1} = 1 & \text{(R1)}, \\ \text{else if} & (N_t \geq E_b) \,\&\, (N_t \leq E_h) \rightarrow S_{t+1} = S_t & \text{(R2)}, \\ \text{else} & S_{t+1} = 0 & \text{(R3)}. \end{array}$$

A systematic study of the set of two-dimensional CA based on this definition allowed us to discover a large number of CA that exhibit the entire spectrum of dynamical behaviors [Heudin 1997]. The most widely used CA classification on the basis of their dynamical behaviors is due to Stephen Wolfram [Wolfram 1984]. He proposed four classes and their analog in the field of dynamical systems:

- Class I CA: evolve to a fixed and homogeneous state (limit point).
- Class II CA: evolve to simple periodic configurations (limit cycles).
- Class III CA: evolve to chaotic patterns (chaotic behavior).
- Class IV CA: evolve to complex patterns of localized structures with long transients (no direct analog).

In the next sections, we will describes Life examples of each class. We based our experiments on a toroïdal universe, consisting of a 64 × 64 × 64 lattice. Each primordial soup experiment was initialized randomly to a 50% density of living cells. Such disordered configurations are typical members of the set of all possible configurations and, therefore, patterns generated from them are typical of those obtained with any initial state. Thus, the presence of structures in these patterns is an indication of self-organization in the CA [Wolfram 1984].

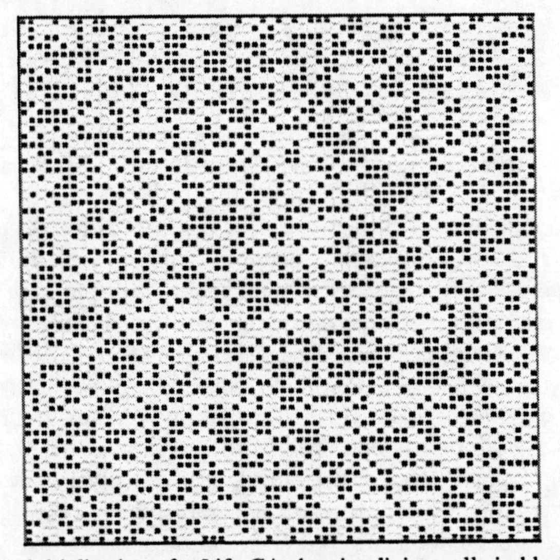

Figure 3. A random initialization of a Life CA showing living cells in black and dead cells in white.

1.3.2. Ordered Dynamics

Both class I, that is fixed-point, and class II Life CA, that is limit cycles, show ordered dynamical behaviors. With a class I CA, any possible initial configuration evolve to a fixed and homogeneous set of living or dead cells. For example, this is the case for Life 5566. After a small number of cycles, called also generations, it stabilizes displaying an array of dead cells. The area of dynamical activity has collapsed to nothing.

There is a large number of class II CA in the Life set, with varying periods from one to large ones. Their dynamics reach a heterogeneous configuration consisting of both types of cells. Even if the period is only one, patterns of final configurations are dependent on the initial states (unlike class I CA). The following figures show two examples of such periodic Life CA. The first one is Life 1444 (cf. figure 4).

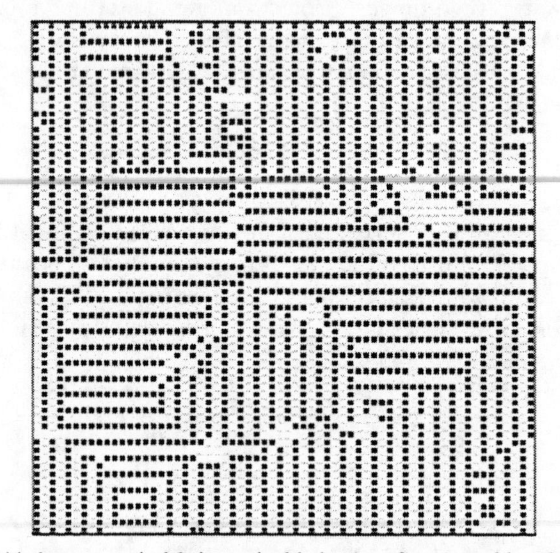

Figure 4. Life 1444 shows a period 2 dynamical behavior after several hundreds generations.

Life 1444 stabilizes with a nice and very ordered period 2 configuration after a transient length of several hundreds generations. During these cycles, it displays horizontal and vertical regions that seem to compete in order to gain more space. When the last configuration is obtained, the lattice is divided into large areas with some small empty "battle field" regions. These areas are the result of the long transients of dynamical activity that collapse down onto isolated periodic patterns.

The second example of class II Life CA is Life 5616 (cf. figure 5). Even if it is a periodic CA like the previous one, Life 5616 is characterized by a slightly different behavior. In a first phase, black and white connected regions grow in the lattice. Then, these regions seem to blink between a set of complementary configurations.

We must note here that the size of the array has an effect on the dynamic of a Life CA. Generally, small arrays have no long transient and fall rapidly onto fixed or short periodic cycles since the behavior of such CA are then totally dominated by boundary effects.

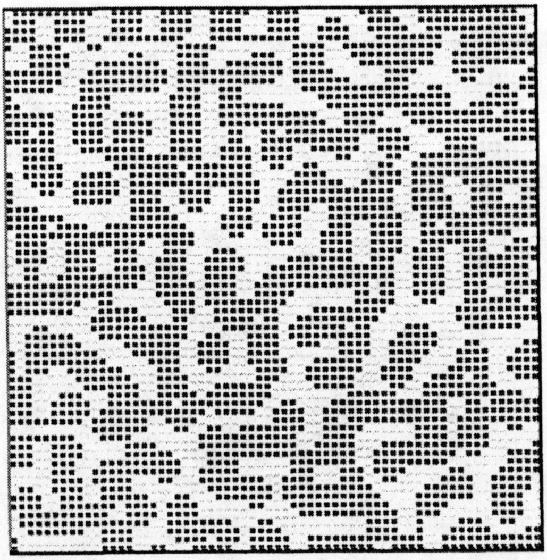

Figure 5. One of the pattern of a final periodic configuration produced by Life 5616 showing the black and white connected regions in the lattice.

1.3.3. Chaotic Dynamics

Some of the Life CA evolve to periodic behavior, but some others settle down to chaotic behavior. Class III CA can exhibit maximal disorder which exists on both local and global scales. This is the case for Life 1122 (cf. figure 6) for which the chaotic behavior is achieved in only a few time steps.

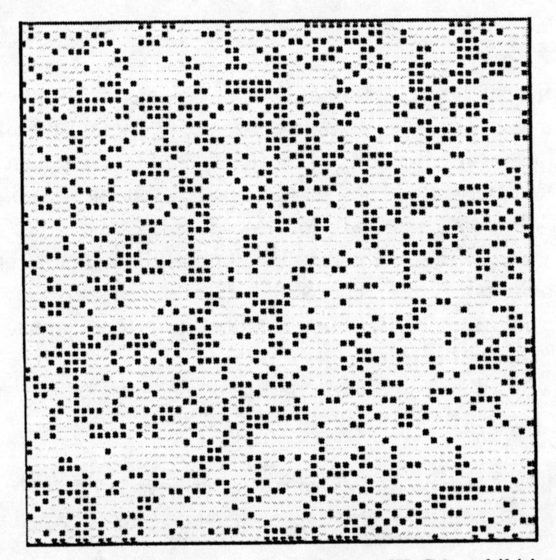

Figure 6. Life 1122 is one of the large number of class III CA exhibiting a fully developed chaotic behavior.

The next figure shows another example of a class III Life CA. Life 2234 is characterized by a global chaotic behavior but with a higher local density of living cells. It shows some local structures that always disappear after a few time steps. This can be interpreted as if the dynamics is able to support the preservation of information for a very small number of cycles before being destroyed by the fluctuation of the environment (cf. figure 7).

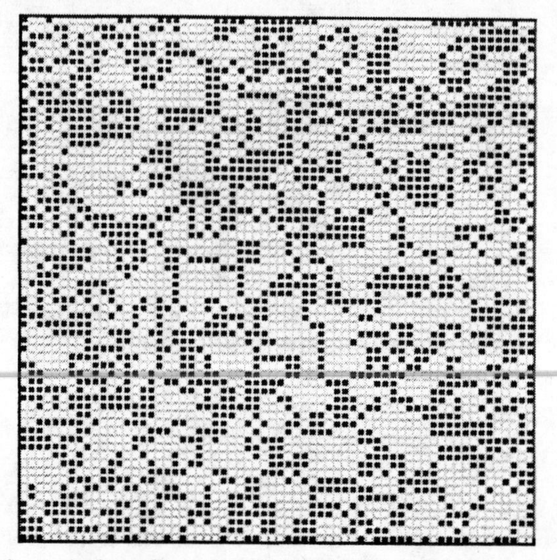

Figure 7. One of the typical configuration of Life 2234 showing local aggregates of living cells.

1.3.4. Complex Dynamics

Among the different Life CA classes, class IV is the most interesting because it includes CA characterized by a "complex" behavior. The most well-known example of such a CA is the so-called Game of Life discovered by John Conway. Conway's Life shows a large number of fixed and periodic patterns along with propagating structures called "gliders". In Bay's notation, Conway's CA is Life 2333 (cf. figure 8). Wolfram suggested that class IV CA are capable of supporting computation and some of them universal computation. Life 2333 has already been shown to be capable of universal computation [Berlekamp 1982] but until recent time, there was no other example of a Life CA with a similar behavior.

Carter Bays has defined two criteria that a CA must satisfy to be a Game of Life "worthy of the name" [Bays 1992]:

1. Primordial soup experiments must exhibit bounded growth. This means than an initial random "blob" must eventually stabilize and cannot grow without limit.

2. A glider must exist and occur "naturally", that is, it must be discoverable by repeatedly performing primordial soup experiments.

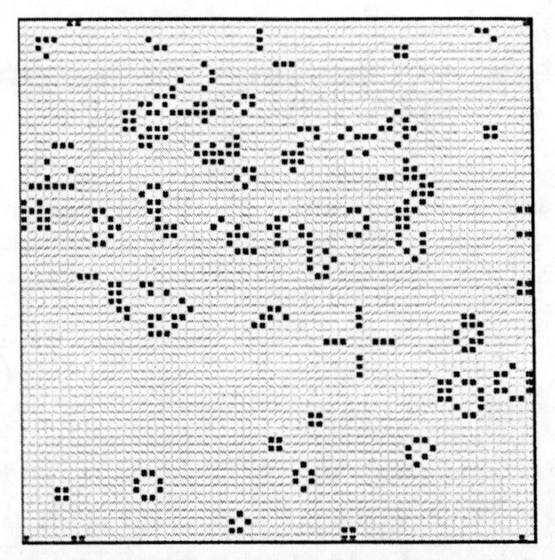

Figure 8. Conway's Life (Life 2333) is the most fascinating CA. Any initial random configuration tend to fragment into a constellation of geometric patterns that grow or flicker out of existence like bacteria in a culture.

While performing our systematic study of the Life CA space, another rule candidate has been discovered [Heudin 1996]. This rule, called Life 1133, satisfies Bays's criteria. The first one is satisfied since all primordial soup experiments we have performed shrunk and stabilized or disappeared. The second criterion is also satisfied since we discovered a nice propagating glider that occurs naturally.

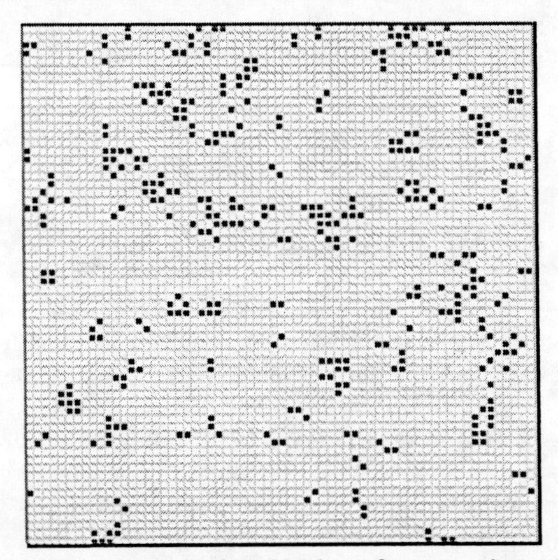

Figure 9. A typical configuration of Life 1133 after a few generations with some fixed and periodic structures.

1.3.5. The Life 1133 World

The global behavior of Life 1133 looks like that of Life 2333, but with shorter transients, that is about 50 generations to be compared with more than 1000 generations for Life 2333. For almost all initial disordered configurations, the density decreases rapidly and tends to an equilibrium limit. In all cases, most sites are seen to "die" after a small finite time. However, stable or periodic patterns which persist for an infinite time are generally formed and, in a few cases, propagating structures have emerged. The low density of Life 1133 after a few time steps is compensated for by the rich variety and symmetry of its Life forms.

The first kind of natural structure includes stable patterns that never change. They represent the commonest objects of Life 1133. Figure 10 gives a selection of these structures, including the "block": a four-cell pattern formed by a two-by-two square which is one of the few structures in common with Life 2333. The most common object of Life 1133 is the "small block" consisting of two adjacent living cells.

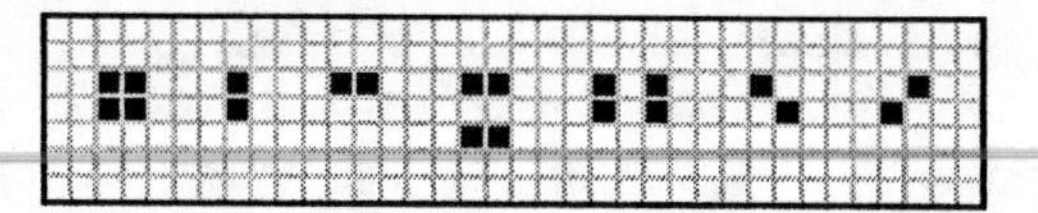

Figure 10. A selection of stable structures including the "block" and many variants of the "small block" made of two adjacent living cells.

Life 1133 shows a rich variety of oscillators. Most of them have a period of two. Figure 11 gives a selection of these objects. The fourth oscillator in the upper part of the figure lead to variants, some of them occurring naturally, such as the first oscillator in the lower part of the figure. One of the most interesting features of Life 1133 seems to be its symmetry property. All natural structures that have been discovered so far exhibit symmetry of one form or another: reflection, rotation, etc.

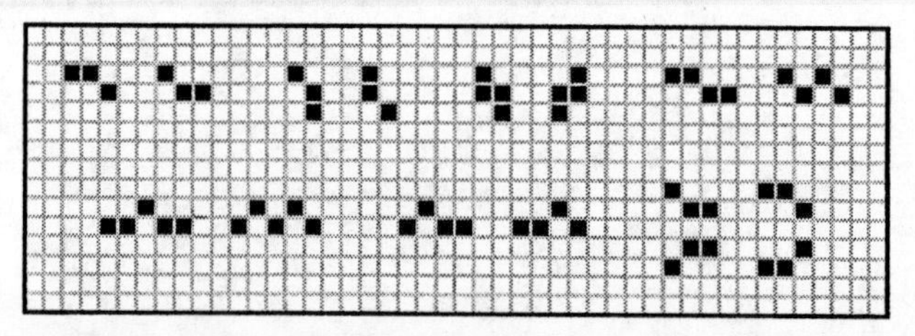

Figure 11. A selection of period 2 oscillators. Many of them are symmetrical variations of the same pattern. For each structure, we give its two phases (right and left patterns).

Oscillators with periods other than two have been found, but all with an even period. Figure 12 shows a selection of period four oscillators. The first one is the "blinker" which looks like the one from Life 2333 but with a period four instead of two and a more sophisticated pattern. The second one, called the "tumbler", consists of four patterns of eight living cells. The third and fourth ones are respectively named the "twister" and the "rattle". When generated on a high-speed computer, these two oscillators create convincing illusions of three-dimensional rotary motion.

One of the most beautiful and interesting objects is a symmetric combination of four "blinkers". This arrangement recalls the well known "traffic light" from Life 2333, (cf. figure 13). With a fast execution speed, this structure looks like a "twinkling star ".

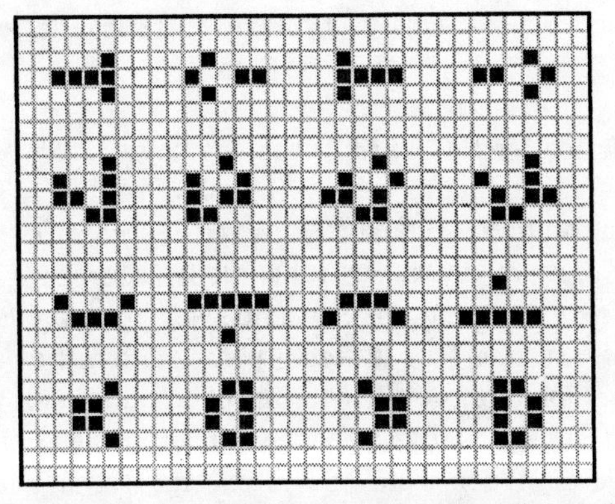

Figure 12. A selection of period 4 oscillators including the blinker (in the upper part), the tumbler, the twister, and the rattle. For each structure, we give its four phases.

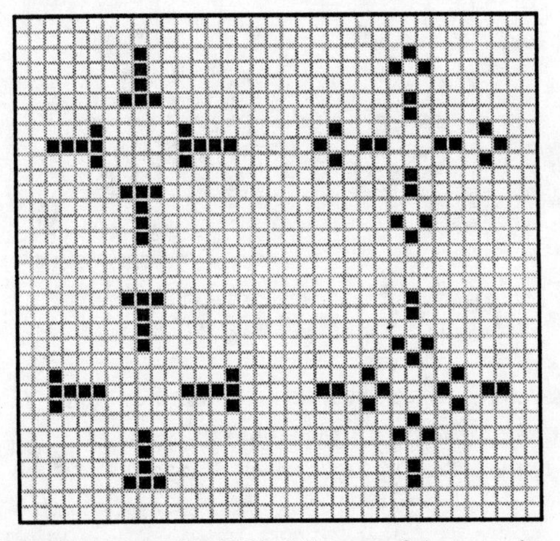

Figure 13. The twinkling star from Life 1133. It is one of the most beautiful pattern we have discovered, but only one time in a very large number of experiments.

A quite common oscillator has a period of ten. We have called it the "alien clock" because of its decimal period and the shape of its patterns (cf. figure 14). With due attention to details, its second five phases are the rotated and mirrored images of its first five phases. Oscillators such as the "alien clock" contribute to the complex kinetic effects displayed by Life 1133 onto a computer screen.

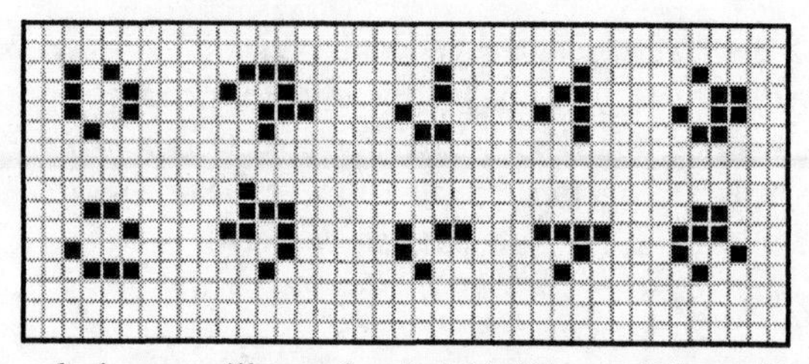

Figure 14. The alien clock, an oscillator of period 10, is one of the emblematic figure of Life 1133.

One of the criterion for Life 1133 to be a Game of Life worthy of the name is the existence of a natural propagating glider. The glider of Life 1133 was discovered only after ten primordial soup experiments. However, it seems that it is less common than the one of Life 2333, mainly because of the relative short transients of Life 1133. The glider appears to creep something like an amoebae, changing its shape as it goes. Like Conway's glider, it assumes four different phases and moves at the same speed: one cell per four generations, or, in other terms, one-quarter cell per generation. Unlike Conway's glider, it moves horizontally or vertically instead of diagonally. Figure 15 shows the four states. When state one is encountered again, the glider will have moved one cell forward.

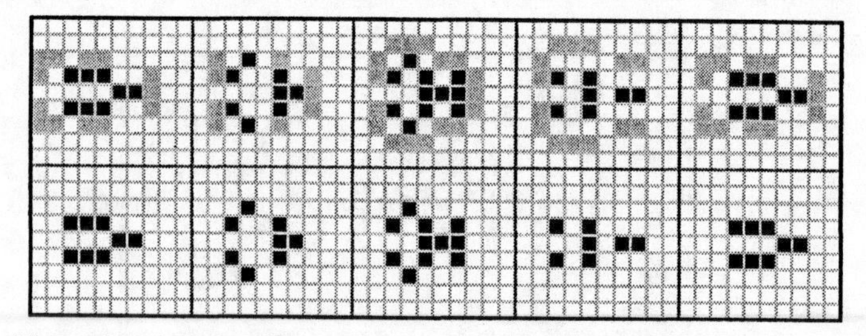

Figure 15. The period 4 glider of Life 1133. The upper part shows the effect of the transition rule (R1: black "on" cells, R2: dark gray "on" cells and gray "off" cells, R3: white "off" cells).

What happens when a glider collides with another object ? The number of possible objects leads to a large number of possible collisions between a glider and other objects. One would expect that, in most cases, the result of such events is the annihilation of both objects. However, in some other cases, the collision leads to the birth of structures that do not fade instantly. The behavior of gliders interacting with other artifacts represents the foundation for implementing complicated structures. One of the most interesting collisions occurs when a glider collides with a small block. This configuration produces a new glider type consisting of both objects. We have named this glider the "fighter", because of its robustness and the fact that it can fire a "torpedo" on a line of small blocks and destroy any object placed at the end of the line (cf. figure 16). Besides this amusing comparison, it gives a clear example of properties of Life 1133 for constructing structures showing complicated or surprising behaviors.

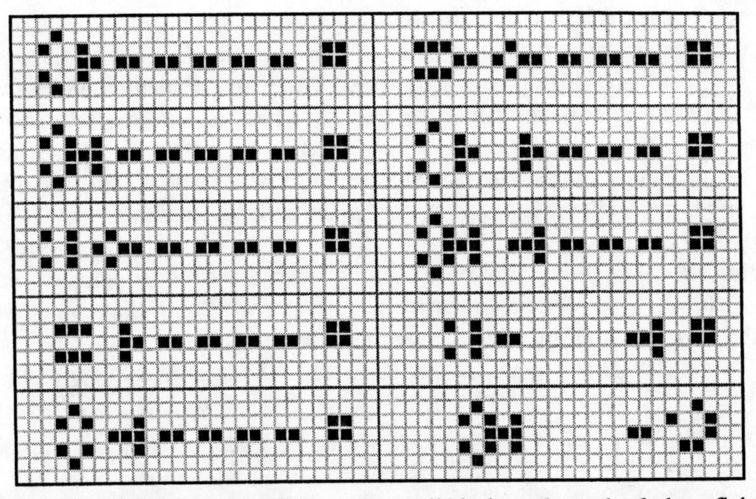

Figure 16. The collision between a glider and a small block produces the 6-phase fighter. This figure shows a 6-phase fighter in action if placed in front of a line of small blocks: time step 1 to 8 (right and upper left), and then 17 and 20 (bottom left).

Life's rules limit the speed of propagating structures. The maximum speed at which any information can be transmitted across the Life plane is one cell per generation in any direction. This can be viewed as the counterpart of the speed of light in the real world and is often called by that name. It is quite easy to find a pattern propagating at this maximum speed in Life 1133. As seen in figure 16 with the fighter, a simple "pulse" on a line of small blocks produces that behavior.

The existence of a glider suggests that, similar to Conway's Life, Life 1133 could be capable of universal computation. In a constructive proof of universality [Berlekamp 1982], a stream of regularly spaced propagating gliders represents a string of bits. In such a stream, the presence of a glider at a given position represents a "1" and absence represents a "0." Based on such streams, the implementation of a single logical gate (a NAND or a NOR gate) is required to be able to design any digital machine, including a universal Turing machine. In practice, there are further elements required such as elements that will turn a signal stream by right angles and a vanish reaction, but we do not go into the details of these elements here. A series of experiments was performed in order to find a "glider gun": a periodic structure that produces the required steady stream of gliders. We have found many configurations producing a glider, but we must point out that we have not yet discovered this crucial element. Note that Carter Bays has not found such object, or at least reported one, during several years of extensive research on three-dimensional Life CA. However, the study of Life 1133 is still recent and is currently undergoing intense investigation.

1.3.6. Between Order and Chaos

With the Life CA set, we have discovered two fundamental phases: ordered and chaotic. As suggested by Christopher Langton [Langton 91], the primarily states of matter, that is solid and fluid, represent similar examples of dynamical behaviors separated by a phase transition. Near this critical phase transition, between too much order and too much chaos, some systems maintain themselves on extended transients. This is the case for living organisms which evolve using the information processing capacity of DNA. In order to compute, a system must be able to store and transmit information and this tradeoff seems to be optimized in the vicinity of a phase transition. The Life CA space exhibit similar dynamical properties with fixed, periodic, chaotic and complex regimes. Thus it provides an interesting framework for studying the evolution of complexity [Heudin 1998a].

With the Life CA, we are experimenting with a sort of "programmable matter" [Toffoli 1987], a simple two-dimensional world for which we can tune physical laws in order to explore the physics of possible worlds. And what we learn about these possible worlds is likely to tell us something about the dynamics of our real world.

1.4. Example of a Three-dimensional Virtual World

1.4.1. The Nut Virtual World

In this section, we describe a second example of a virtual world but rather based on a 3D immersive environment. The aim of this ongoing project is to synthesize a 3D universe on the internet for studying artificial evolution of digital creatures. The name Nut recalls the Egyptian sky goddess. She was depicted as a naked giant woman who was supporting the sky with her back. Ancient documents describe how each evening, the Sun entered the mouth of Nut and passing through her body was born each morning coming out of her womb.

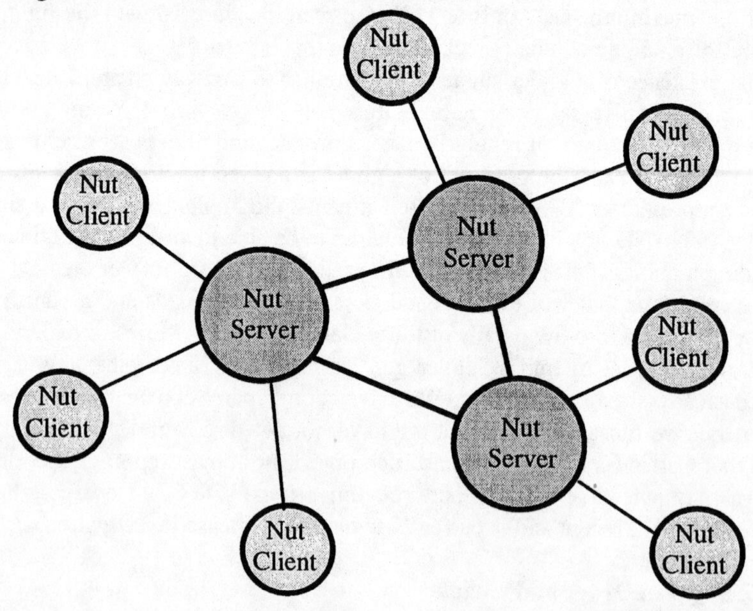

Figure 17. Nut is based on a many-server many-client architecture. Each server manages one or more virtual environments connected together via communication channels.

1.4.2. Nut Architecture Overview

The Nut virtual world is based on a many-server many-client software architecture (cf. figure 17). This architecture uses the standard Internet technology and protocols, and thus does not require any dedicated hardware. Nut allows experiments with small configurations like a single personal computer to large simulations using many servers and clients connected together and forming a huge virtual environment. The virtual space of each server is connected to one or more other spaces through "teleporting" channels. At a given time, a client is linked to only one server, but the user can move from space to space by entering these communication channels. This results in a move from one server to another server.

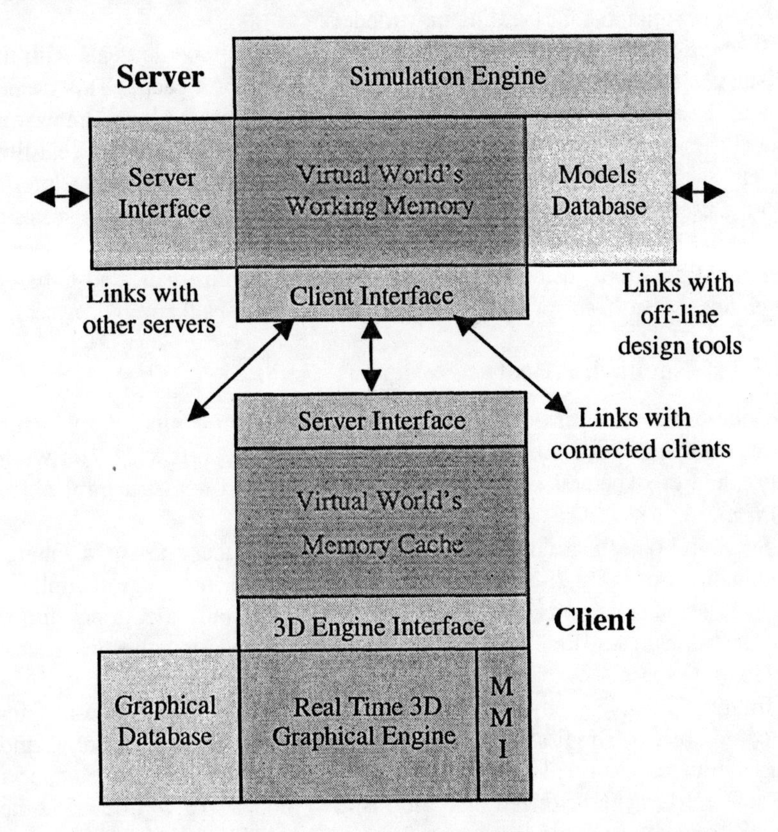

Figure 18. Diagram of Nut's basic software showing the software organization of a server and a client.

The Nut virtual world architecture makes a clear distinction between the simulation engines and the 3D graphical rendering engines (cf. figure 18). The simulation engine deals only with environmental and creature and is located on servers. The graphical and navigation software represents the client part of the architecture. This allows to use 3D rendering engines which can be based on various technologies depending of the nature of the experiment and/or of the user's hardware environment.

Information exchanges between servers and clients occur at the simulation level during a run. Graphical information like textures are locals to the client and can be transmitted before or between experiments.

The basic architecture of a server includes a simulation engine that deals with the 3D world model located in the working memory. The engine simulates "physical laws" such as gravitation and "biological laws" such as Darwinian evolution. The simulation engine is implemented using the multi-threading capabilities available on most server's operating systems. The client interface manages connections and communications with clients. A database for models enable to store on a hard disk all the models required by the simulation engine. It also constitutes a good interface with off-the-shelf design tools for creating new models.

Each client includes at least two modules. The first module deals with the server. It manages connections and maintains the coherence of a cache. This cache stores a copy of the server's working memory, the one that is accessible by the user at a given time. The second module is the 3D graphical engine displaying in real-time all the models of the user's point of view in the cache. The Man-Machine Interface deals with the user's command and control actions. The graphical database stores all necessary rendering data such as textures. The 3D engine interface allows to design various client configurations, using different technologies such as dedicated C-based rendering engines, Java-based or VRML-based implementations for web browsers.

1.4.3. The Nut Environment

The Nut virtual world is a full 3D dynamic environment composed of a set of worlds connected through communication channels called "teleporters". A Nut world is basically made of structural objects grouped in a *map* and non-structural objects called *entities*.

Structural objects are individual 3D shapes like a cube, a prism, a sphere, etc, and combinations of these simple objects can be arranged into more complicated structures. As standalone units, they can be reshaped into walls, floors or ceilings, bodies of water, skies, etc. Then, all these units can be grouped together in order to form a coherent 3D space.

Entities include "physical" non-structural objects and "non-physical" non-structural objects. The first class includes animated objects that can move and respond to user-initiated events like a door for example. *Creatures* represent a special case of physical entities characterized by a bio-inspired model. We will describe them in the next sections. The second class includes insubstantial and invisible objects like triggers, light sources and sound sources. Every entity has one or more properties like behavioral information and collision detection. Other examples are the brightness of a light source, the sound of a sound entity, the speed of a moving object, or the mass of an object for gravity simulation.

1.4.4. Evolution of Digital Creatures

One of the most remarkable fact about biological evolution is that the co-evolutionary process has produced more and more sophisticated organisms with powerful cognitive capabilities. The main goal of the Nut virtual world is to study the evolution of digital creatures. The application of an artificial evolution process will lead to an open-ended power and will generate surprising consequences.

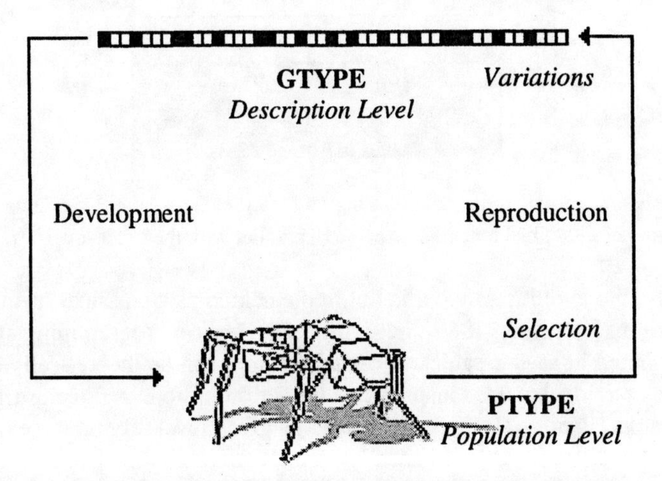

Figure 19. This simplified diagram shows the relation between the generalized genotype, that is the encoded self-description of the creature, and the generalized phenotype, that is the resulting individual in the virtual environment.

One of the most salient and universal characteristic of living organisms, from an organizational point of view, is the distinction which has been made by biologists between the genotype (the genetic code made of DNA chains) and the phenotype (the organism itself). As initially proposed by Christopher Langton [Langton 1988], we must refer to generalized notions so that we may apply them in a non-biological situation like the Nut virtual world. Thus, we define the generalized genotype, called GTYPE, to refer to the self-description of a creature. We define also the generalized phenotype, called PTYPE, to refer to the set of structures and behaviors that emerges out the interpretation of the GTYPE in the virtual environment.

These two levels are linked together by the two main processes of the Nut artificial evolution scheme. The first process is to the morphological *development* phase of a creature. It creates the PTYPE of a creature by interpreting its GTYPE in the virtual environment, generating its 3D model and associated dynamics. The second process is the *reproduction* phase. Its creates a new GTYPE by applying variation operators, like mutation and crossing-over, on a parent GTYPE. These two phases are parts of a Darwinian-like evolution cycle which can be briefly described as follow. A set of GTYPEs generates a population of PTYPEs which interacts within the virtual environment. On the basis of their relative performance, some of these PTYPEs "survive" and reproduce, creating new GTYPEs. Then, a new generation of PTYPEs emerges out the interpretation of their respective GTYPE and this process continues. Evolution works by selecting the fittest PTYPEs among variants.

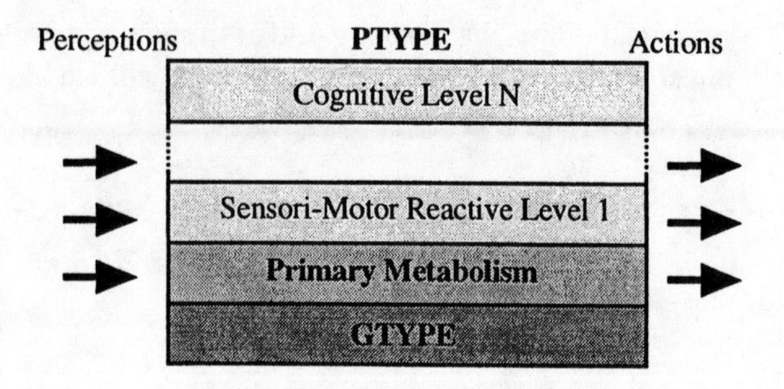

Figure 20. Theoretical layered model of a digital creature. Each level represents a set of "perception-computation-action" units that execute in parallel with the other levels.

This artificial evolution model is not implemented as a classical (centralized) Genetic Algorithm [Goldberg 1989], but rather as a Genetic Programming system [Koza 1992] distributed in each creature. A digital creature can be theoretically described as a layered hierarchical model inspired by the subsumption architecture proposed by Rodney Brooks [Brooks 1991]. In this model, a given level rely on the existence of its sub-levels and all levels are intrinsically parallel. Unlike the subsumption model, the two elementary levels of each creature are its GTYPE and a *Primary Metabolism.* In biology, a metabolism is responsible for taking material from the environment and transforming it in order to create the form of the organism. The Primary Metabolism we refer to can be described as a recursive partially stochastic procedure responsible for self-creation (development), self-perpetuation (survivance) and self-reproduction in the meaning of the autopoesis model proposed by Varela [Varela 1989]. This three functions are strongly linked together and form the primary dynamics of any creature. All other levels, including sensori-motor and cognitive levels (if any) rely on the existence of these two basic organizational levels.

1.4.5. Generalized Genotype

A GTYPE is the complete self-description of a digital creature. It can be viewed as a string of parameters that will be used by the Primary Metabolism when executing its main functions that is morphological development, evaluation of fitness for survivance, and self-reproduction. It also includes some low-level executable code like the Primary Metabolism procedures themselves.

Genotypes used in Artificial Evolution and Genetic Algorithms consist generally of fixed-size strings of binary informations. Such fixed-size encoding is required in applications like function optimization or problem solving for convergence, but inappropriate when considering the evolution of digital creatures from simple organisms to more complex ones [Harvey 1994]. We use a variable length GTYPE based on a hierarchical representation inspired by the Genetic Programming approach [Koza 1992] and Karl Sims's nested graphs [Sims 1995]. It can be described as a directed graph made of connections and nodes. This graph can be linear or be recurrent forming recursive or fractal-like structures.

A typical connection includes informations like an iteration value for limiting the number of loops when generating recursive forms or a terminal flag that indicates an ending part.

A node describes a part of the creature and includes instructions for growing its morphology and behaviors. The root-node is the starting node of any creature. It contains informations and codes about the Primary Metabolism, and also global parameters such as macro-structural genes (symmetries, segmentation, maximum iteration level, etc) and control genes (mutation probability, crossing-over probability, etc). Each of the other nodes describes a segment of the creature. It contains informations like the type of the segment (head, body segment, leg, etc), dimensions that determine its physical shape, and joint informations about the type of connections with the other segments.

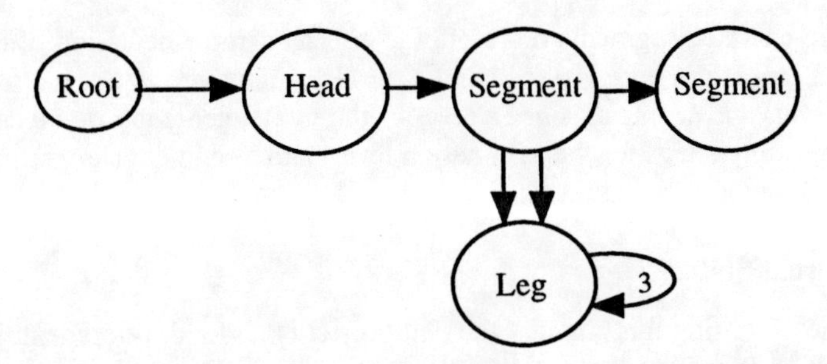

Figure 21. Graph representation of a GTYPE. This diagram shows the basic organization of an insect-like creature.

Each node includes also a sub-graph describing the control part of the segment. This nested graph is a hierarchical description of the perception-computation-action unit located in the segment. It is encoded as a graph of nodes and connections using the same model as used for the creature's morphology. However, in this case, nodes represent elementary control building blocks. There are three main types of such building blocks: sensors (contact sensor, visual sensor, etc), operators (sum, neuron, sinwave generator, memory cell, etc) and effectors (joint effector, sound generator, etc). All these building blocks constitute the basic set of elements that defines the genetic programming language of the Nut system. It is the counterpart of the set of functions and variables defined in the Genetic Programming approach [Koza 1992].

1.4.6. Generalized Phenotype

The PTYPE of a digital creature is the hierarchy of articulated 3D rigid segments that emerges out the development of the GTYPE by the Primary Metabolism. When a creature is synthesized, the root-node of its GTYPE is first activated. Then, all segments described in the GTYPE are recursively generated including their control parts. This causes segments to be created or duplicated as identical parts when required. Note that the development process can include stochastic alteration in order to have more realistic forms [Oppenheimer 1988].

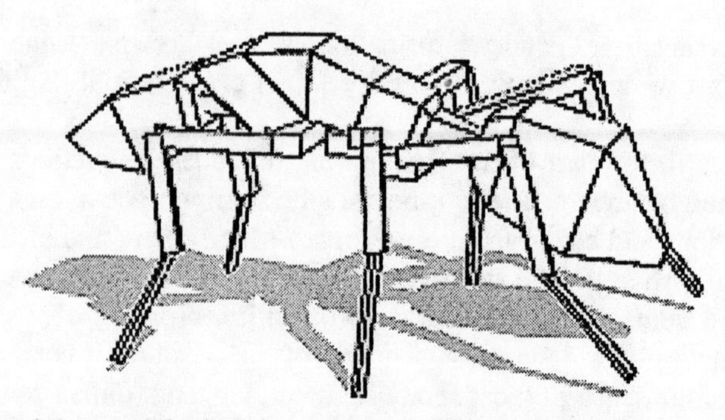

Figure 22. One of the possible digital creature resulting from the development of the insect-like GTYPE shown in figure 21.

All the generated segments represent the effective implementation of the theoretical layered model presented in section 1.4.4. Note that there is not a direct mapping between levels and segments since a level of the model generally involves many segments (for example legs for the locomotion level) and a segment can participate in the implementation of different levels.

1.4.7. Current Status

At the time of writing this chapter, the Nut project is under development. It is consistent with Karl Sims's experiments on the evolution of 3D morphology and behavior by competition [Sims 1995] which was implemented to run on a CM-5 parallel computer and required several hours of computing times. The first implementation of Nut uses a C-based dedicated 3D engine running on a Linux personal computer. It enables to display in real-time large virtual spaces inhabited by evolved digital creatures. First creature populations evolve not from an initial random population as in Sims's study, but rather from pre-designed artificial life forms like the insect creature shown in the previous figures.

1.5. Philosophical Implications

1.5.1. Are these Virtual Worlds Real ?

The synthesis of virtual worlds raises many fundamental questions about the nature of reality, virtuality and life. In this section, we will not try to make an exhaustive list of these questions but rather we will introduce some of them in order to open the debate. One important philosophical question concerns the ontological status of virtual worlds. Is a virtual world a real world ?

In the 1950's and 1960's, the mind-body problem has witnessed a succession of theories: from logical behaviorism [Ryle 1949][Wittgenstein 1953] to materialism with the identity theory [Place 1956][Smart 1959]. Functionalism has been also a major school of thought in the philosophical debate on the ontological status of intelligence and consciousness [Putnam 1975][Fodor 1968][Dennett 1978].

The debate about the reality of virtual worlds has something to share with the debate about the possibility of machine intelligence. Like Artificial Intelligence (AI) and AL, VW study its subject matter by attempting to realize it within computers. The position that a world's processes can be abstracted from matter can be therefore considered as a functionalist position. Thus, as for AI and AL, there are two claims that can be made: a *weak* claim and a *strong* claim (by weak we mean here "modest" and by strong we mean "daring").

The weak claim holds that all these computer programs are nothing more than simulations. These simulations are very useful for addressing scientific questions about reality but they could never be considered to be instances of worlds themselves. They are just ungrounded symbol systems that are systematically interpretable as if they were real but they are not.

The strong claim holds that some of these virtual worlds could be as real as our physical world is. In a sense, a virtual world is very physical: it exists within a computer made of physical matter, it consumes energy and produces heat. A virtual world is a computer program made of information flows composed of electrons traveling within integrated circuits and through connections.

1.5.2. Virtuality vs. Reality

A virtual object is classically defined as an object existing within an optical system, that is composed of a source of light and an observer, but with no physical reality. In the context of VW, virtual means purely computational, that is made of computer programs.

Reality is a physical process. Thanks to its universality, it is possible to simulate any process on a computer. Hence from this two postulates, it should be possible to simulate the real world on a universal computer. This simulated world would be an artificial system, that is a set of symbols that imitates real objects by an attempt to capture some of their properties or behaviors. There may be a one-to-one correspondence between the objects and their symbolic descriptions. However, the simulation of these objects is not the same as the real objects. In the case of the real world, objects are what they are intrinsically, without the mediation of any interpretation. In the case of the simulation, the only objects are the symbols and their syntactically constrained relations. The rest is just our interpretation of the symbols and their interactions *as if* they were properties of the objects they describe.

The problem becomes more complex if we consider the synthesis of an imaginary virtual world with its own "physical" and "biological" laws. Imagine that this world includes artificial organisms perceiving their "reality", which for them, could be as real as our reality is for us. In this case, there is no direct correspondence between our reality and the virtual world. The virtual world exists in a computer and its properties may be very different of the one of our real world. The basic idea behind this concept is that this virtual world could be a self-synthesized system of existence [Wheeler1988]. In this framework, Rasmussen argued that a reality obtains its meaning through the existence of a living observer [Rasmussen 1991]. Since the virtual world is being perceived by an observer, just as our world is being perceived, it becomes a reality with an equal ontological status. We will not address here all the philosophical questions raised by the possibility of artificial organisms, since it has been largely discussed in the past [Sober 1991][Keeley 1994][Harnad 1994][Emmeche 1994]. We will only suggest that this living observer could be an autonomous digital creature or a human's avatar immersed in the virtual world. Therefore, a *real* world in a computer should, thereby, be possible.

1.5.3. The Dream of Reality

Virtual Worlds forces us to rethink what is to be "real". It is very difficult to set up objective criteria by which we can define reality. The common sense view of our physical world is that there is an absolute physical reality independent of us. The geometry of this reality is a 3D space in which all the events occur. This is often called the Newtonian world. However, we know from modern physics that matter and geometry are coupled. It is impossible to define one without the other. Our ordinary notions of space and time are just a convenient fiction.

The constructivist approach has suggested that the world cannot be divided into two separate realities: on one hand the subjective world of our sensory experience, and on the other hand, the objective world of reality. We cannot access a reality independently of our perception. Thus, we could say that we construct or invent the reality rather than we discover it [Segal 1986]. It does not means that we imagine a world that does not exits, because this extreme position would result in a solipsist view [Joad 1936]. There is a real world independent of our act to know, but this world is not directly accessible.

The traditional approach of science has been to reduce the reality to a set of elementary particles. It then explains sensory phenomena, such as color or warmth, in term of the spacetime patterns of these particles. However, reality is indescribably more rich and complex. There is a large number of complexity levels which are linked to our perception [Heudin 1998a]. We perceive the reality through our senses, and this is true for the physical world as well as for virtual worlds.

1.6. Virtual Worlds and Art

1.6.1. Relationships with Other Art Forms

Nothing can reach out and grab your attention like a book can. You can read it from the first word to the last, or you can shuffle between multiple passages, inserting bookmarks to mark your way. With a book, there is a high degree of freedom for interpreting the text. The resulting abstract world is built not only on this text but also on our own history and sensory experiences.

A movie can transport you to a place and time, or immerse you in an imaginary space inhabited by terrifying aliens. We sit in the audience and watch the action taking place on the screen. All is carefully designed to deliver an aggregate interpretation of the author, director and actors.

Despite their evidente differences, books and movies propose experiences which are similar to the one of Virtual Worlds, submerging their audiences into the plenty of other universes. Most of the differences lies in their respective degrees of *subjective presence* (an objective presence would be to be physically present) and *interactivity*.

We can extend this idea to artworks in general. We usually think of an artwork as a physical object. However, an artwork also lies in the concepts and feelings that are suggested and produced by our intellect. Artworks create immersive virtual worlds but most of them are non-interactive. Early computer art was based on batch-processing. Now, with the advances in computer technology, artists can create new art forms based on interactivity (cf. Chapter 10). VW could be on of these new forms, a major new medium based on the collaboration of science, technology and art.

1.6.2. Science and Art

As we emphasized, VW is concerned with using the computer as a tool to help understand the universe around us. It is like a telescope or a microscope, allowing us to see the unseable. The sciences of complexity are still very much in its infancy and many of underlying ideas were anticipated by generative art. The early form of computer art were almost entirely exploration of the computer's capacity for generating complex images. Many pieces have explored the theme of how complexity grows from the working of simple rules. As we look back to them, we see now illustrations of the new fractal and chaos theories [Mallen 1993].

Art exists thus as the counterweight and the complement of science, so that the two together complete the human discovery process [Kisseleva 1998]. Most of the time, scientists seeking to explain or understand complexity have to create a new visual vocabulary to do this. This is an imaginative and aesthetic act which emphasizes that the creative and understanding acts are inseparable. This is one of the main reason why art will play an important role in the new field of VW.

1.7. The Coming Evolution

For many years, AI was supposedly concerned with generating intelligent behaviors. Researchers have tried to create artificial brains but with weak results compared to early predictions. One of the main reasons is that the brain is one of the most complex system we know and, unfortunately, even today, underlying mechanisms are not well known. Another important reason is that it is an intrinsic part of an individual which is the result of millions years of evolution. Such a complex evolved system cannot be designed *ex nihilo* and separated from the rest of an organism conceived as a whole.

Since 1987, AL is concerned with the study of life and evolution. AL researchers modelize the evolution and behaviors of very simple organisms. Since the focus of AL is the study of artificial living organisms, most simulated environments are very simple, providing the basic resources that are necessary for the experiment. However, any biologist knows that species are the products of their environment, including the complex co-evolutionary processes that occur within this environment. Like a brain that cannot be separated from its body, a living creature cannot be separated from its environment.

The next coming evolution will be the synthesis of artificial worlds as wholes. By attempting to synthesize complex digital environments, including their physical and biological laws, Virtual Worlds represent a new trend of research in the sciences of complexity.

The study of virtual worlds may potentially address very important questions about complexity and reality. It also provide a new forum to address some of the most fundamental philosophical questions. Besides scientific inquiry, philosophical debates and art, why should research in this area be supported ?

There are many reasons for doing so, but one of the most important is the possible outcomes in terms of applications. These applications range from computer games, simulation of complex environments, virtual universities and virtual shopping malls on the internet, virtual galleries and museums, and more, to new applications that we have never think of before.

References

Adami, C., Brown, C.T., 1994, Evolutionary Learning in the 2D Artificial Life System "Avida", *Artificial Life IV*, Proceedings of the Fourth International Workshop on the Synthesis and Simulation of Living Systems, edited by R.A. Brooks and P. Maes, MIT Press (Cambridge), 377.

Bays, C., 1987, Candidates for the Game of Life in Three Dimensions, *Complex Systems*, **1**, 373.

Bays, C., 1988, A Note on the Discovery of a New Game of Three-Dimensional Life, *Complex Systems*, **2**, 255.

Bays, C., 1990, *The Discovery of a New Glider for the Game of Three-dimensional Life, Complex Systems*, **4**, 599.

Bays, C., 1992, A New Candidate Rule for the Game of Three-dimensional Life, *Complex Systems*, **6**, 433.

Berlekamp, E., Conway, J.H., Guy, R., 1982, *Winning Ways for Your Mathematical Plays*, Academic Press (New York).

Brooks, R., 1991, Challenges for Complete Creature Architectures. *From Animals to Animats*, edited by J.A. Meyer S.W. Wilson, MIT Press (Cambridge).

Brooks, R., Maes, P. (ed.), 1994, *Artificial Life IV*, Proceedings of the Fourth International Workshop on the Synthesis and Simulation of Living Systems, MIT Press (Cambridge).

Dennett, D., 1978, *Brainstorms*, MIT Press (Cambridge).

Emmeche, C., 1994, Is Life as a Multiverse Phenomenon ?, *Artificial Life III*, SFI Studies in the Sciences of Complexity, Addison-Wesley (Reading), **17**, 553.

Fodor, J., 1968, *Psychological Explanation*, Random House (New York).

Gardner, M., 1970, The Fantastic Combinations of John Conway's New Solitaire Game Life, *Scientific American*, **223**, 120.

Goldberg, D.E., 1989, *Genetic Algorithms in Search, Optimization and Machine Learning*, Addison-Wesley (Reading).

Harnad, S., 1994, Artificial Life: Synthetic vs. Virtual, *Artificial Life III*, SFI Studies in the Sciences of Complexity, Addison-Wesley (Reading), **17**, 539.

Hartman, J., Wernecke, J., 1996, *The VRML 2.0 Handbook - Building Moving Worlds on the Web*, Addison-Wesley Developers Press (Reading).

Harvey, I., 1994, Evolutionary Robotics and SAGA: The Case for Hill Crawling and Tournament Selection, *Artificial Life III*, edited by C.G. Langton, SFI Studies in the Sciences of Complexity, Addison-Wesley (Reading), **17**, 299.

Heudin, J.C., 1996, A New Candidate Rule for the Game of Two-dimensional Life, *Complex Systems*, **10**, 367.

Heudin, J.C., 1997, *Complexity Classes in Two-dimensional Life-like Cellular Automata*, IIM Lab. Working Paper.

Heudin, J.C., 1998a, *L'évolution au bord du chaos*, Editions Hermès (Paris).

Heudin, J.C. (ed.), 1998b, *Virtual Worlds - Proceedings of the First Int. Conf. on Virtual Worlds*, Springer-Verlag Lecture Notes in Computer Science (Berlin), **1434**, 5.

Husbands, P., Harvey, I. (ed.), 1997, *Fourth European Conference on Artificial Life*, MIT Press (Cambridge).

Joad, C.E.M., 1936, *Guide to Philosophy*, Dover Publications (New York).

Keeley, B.L., 1994, Against the Global Replacement: On the Application of the Philosophy of Artificial Intelligence to Artificial Life, *Artificial Life III*, SFI Studies in the Sciences of Complexity, Addison-Wesley (Reading), **17**, 569.

Kisseleva, O., 1998, Art and Virtual Worlds, *Virtual Worlds - Proceedings of the First Int. Conf. on Virtual Worlds*, edited by J.C. Heudin, Springer-Verlag Lecture Notes in Computer Science (Berlin), **1434**, 357.

Koza, J.R., 1992, *Genetic Programming – On the Programming of Computers by Means of Natural Selection*, MIT Press (Cambridge).

Krueger, M.W., 1991, *Artificial Reality II*, Addison-Wesley (Reading).

Langton, C.G., 1988, Artificial Life, *Artificial Life*, edited by C.G. Langton, SFI Studies in the Sciences of Complexity, Addison-Wesley (Reading), **6**, 1.

Langton, C.G., 1991, Life at the Edge of Chaos, *Artificial Life II*, SFI Studies in the Sciences of Complexity, edited by C.G. Langton, C. Taylor, J.D. Farmer & S. Rasmussen, Addison-Wesley (Reading), **10**, 41.

Langton, C.G. (ed.), 1994, *Artificial Life III*, SFI Studies in the Sciences of Complexity, Addison-Wesley (Reading), **17**.

Mallen, G., 1993, Back to the Cave – Cultural Perspectives on Virtual Reality, *Virtual Reality Systems*, edited by R.A. Earnshaw, M.A. Gigante and H. Jones, Academic Press (london), 266.

McIntosh, H., 1990, Wolfram's Class IV Automata and a Good Life, *Physica D*, **45**, 105.

Oppenheimer, P., 1988, The Artificial Menagerie, *Artificial Life*, edited by C.G. Langton, SFI Studies in the Sciences of Complexity, Addison-Wesley (Reading), **6**, 251.

Packard, N., Wolfram, S., 1985, Two-Dimensional Cellular Automata, *Journal of Statistical Physics*, **38**, 901.

Place, U.T., 1956, Is Consciousness a Brain Process ?, *Brit. Journal Psychology*, **47**, 44.

Poundstone, W., 1985, *The Recursive Universe*, Contemporary Books (Chicago).

Putnam, H., 1967, The nature of Mental States, *Mind, Language, and Reality*, Cambridge University Press (Cambridge).

Rasmussen, S., 1991, Aspects of Information, Life, Reality, and Physics, *Artificial Life II*, SFI Studies in the Sciences of Complexity, edited by C.G. Langton, C. Taylor, J.D. Farmer, & S. Rasmussen, Addison-Welsey (Reading), **10**, 767.

Ray, T.S., 1991, An Approach to the Syntheis of Life, *Artificial Life II*, SFI Studies in the Sciences of Complexity, edited by C.G. Langton, C. Taylor, J.D. Farmer, & S. Rasmussen, Addison-Welsey (Reading), **10**, 371.

Ryle, G., 1949, *The concept of Mind*, Barnes and Noble (New York).

Sanchez, E., Tomassini, M. (ed.), 1996, *Towards Evolvable Hardware - The Evolutionary Engineering Approach*, Springer-Verlag Lecture Notes in Computer Science (Berlin), **1062**.

Segal, L., 1986, *The Dream of Reality*, Norton & Company (New York).

Sims, K., 1995, Evolving 3D Morphology and Behavior by Competition, *Artificial Life*, MIT Press (Cambridge), **1**, 353.

Smart, J.J.C., 1959, Sensations and Brain Processes, *Philosophy Review*, **68**, 141.

Sober, E., 1991, Learning from Functionalism – Prospects for Strong Artificial Life, *Artificial Life II*, SFI Studies in the Sciences of Complexity, edited by C.G. Langton, C. Taylor, J.D. Farmer & S. Rasmussen, Addison-Wesley (Reading), **10**, 749.

Sutherland, I., 1965, The Ultimate Display, *Proceedings IFIP Congress*, 506.

Terzopoulos, D., Xiaoyuan, T., Radek, G., 1995, Artificial Fishes: Autonomous Locomotion, Perception, Behavior, and Learning in a Simulated Physical World, *Artificial Life*, MIT Press (Cambridge), **1**, 327.

Toffoli, T., Margolus, N., 1987, *Cellular Automata Machines*, MIT Press (Cambridge).

Varela, F., 1989, *Autonomy et connaissance*, Editions du Seuil (Paris).

Ventrella, J., 1998, Designing Emergence in Animated Artificial Life Worlds, *Virtual Worlds - Proceedings of the First Int. Conf. on Virtual Worlds*, edited by J.C. Heudin, Springer-Verlag Lecture Notes in Computer Science (Berlin), **1434**, 143.

Verna, D., Grumbach, A., 1998, Can we define Virtual Reality? The MRIC Model, in *Virtual Worlds*, edited by J.C. Heudin, Springer-Verlag Lecture Notes in Computer Science (Berlin), **1434**, 41.

Von Neumann, J., 1966, *Theory of Self-Reproducing Automata*, edited and completed by A.W. Burks, University of Illinois Press (Urbana).

Wheeler, J.A., 1988, World as System Self-Synthesized by Quantum Networking, *IBM Journal of research and Development*, **32**, 4.

Wittgenstein, L., 1953, *Philosophical Investigations*, Basil Blackwell (Oxford).

Wolfram, S., 1984, Universality and Complexity in Cellular Automata, *Physica D*, **10**, 1.

Yaeger, L., 1994, Computational Genetics, Physiology, Metabolism, Neural Systems, Learning, Vision, and Behavior or PolyWorld: Life in a New Context, *Artificial Life III*, edited by C.G. Langton, SFI Studies in the Sciences of Complexity, Addison-Wesley (Reading), **17**, 263.

Chapter 2

An Evolutionary Approach to Synthetic Biology: Zen and the Art of Creating Life

Thomas S. Ray[1]
ATR Human Information Processing Research Laboratories
ray@hip.atr.co.jp
http://www.hip.atr.co.jp/~ray

2.1. Abstract

Our concepts of biology, evolution and complexity are constrained by having observed only a single instance of life, life on Earth. A truly comparative biology is needed to extend these concepts. Because we can not observe life on other planets, we are left with the alternative of creating artificial life forms on Earth. I will discuss the approach of inoculating evolution by natural selection into the medium of the digital computer. This is not a physical/chemical medium, it is a logical/informational medium. Thus these new instances of evolution are not subject to the same physical laws as organic evolution (e.g., the laws of thermodynamics), and therefore exist in what amounts to another universe, governed by the "physical laws" of the logic of the computer. This exercise gives us a broader perspective on what evolution is and what it does.

An evolutionary approach to synthetic biology consists of inoculating the process of evolution by natural selection into an artificial medium. Evolution is then allowed to find the natural forms of living organisms in the artificial medium. These are not models of life, but independent instances of life. This essay is intended to communicate a way of thinking about synthetic biology that leads to a particular approach: to understand and respect the natural form of the artificial medium, to facilitate the process of evolution in generating forms that are adapted to the medium, and to let evolution find forms and processes that naturally exploit the possibilities inherent in the medium. Examples are cited of synthetic biology embedded in the computational

---

[1] This chapter is based on a paper originally published in 1994 in *Artificial Life*, MIT Press (Cambridge), **1**, 179.

medium, where in addition to being an exercise in experimental comparative evolutionary biology, it is also a possible means of harnessing the evolutionary process for the production of complex computer software.

2.2. Synthetic Biology

In Artificial Life (AL) is the enterprise of understanding biology by constructing biological phenomena out of artificial components, rather than breaking natural life forms down into their component parts. It is the synthetic rather than the reductionist approach. I will describe an approach to the synthesis of artificial living forms that exhibit natural evolution.

The umbrella of Artificial Life is broad, and covers three principal approaches to synthesis: in hardware (e.g., robotics, nanotechnology), in software (e.g., replicating and evolving computer programs), in wetware (e.g., replicating and evolving organic molecules, nucleic acids or others). This essay will focus on software synthesis, a though it is hoped that the issues discussed will be generalizable to any synthesis involving the process of evolution.

I would like to suggest that software syntheses in AL could be divided into two kinds: simulations and instantiations of life processes. AL simulations represent an advance in biological modeling, based on a bottom-up approach, that has been made possible by the increase of available computational power. In the older approaches to modeling of ecological or evolutionary phenomena, systems of differential equations were set up that expressed relationships between covarying quantities of entities (i.e., genes, alleles, individuals, or species) in the populations or communities.

The new bottom up approach creates a population of data structures, with each instance of the data structure corresponding to a single entity. These structures contain variables defining the state of an individual. Rules are defined as to how the individuals interact with one another and with the environment. As the simulation runs, populations of these data structures interact according to local rules, and the global behavior of the system emerges from those interactions. Several very good examples of bottom up ecological models have appeared in the AL literature [Hogeweg 1989][Taylor 1989]. However, ecologists have also developed this same approach independently of the AL movement, and have called the approach "individual based" models [DeAngelis 1992][Huston 1993].

The second approach to software synthesis is what I have called instantiation rather than simulation. In simulation, data structures are created which contain variables that represent the states of the entities being modeled. The important point is that in simulation, the data in the computer is treated as a representation of something else, such as a population of mosquitoes or trees. In instantiation, the data in the computer does not represent anything else. The data patterns in an instantiation are considered to be living forms in their own right, and are not models of any natural life form. These can from the basis of a comparative biology [Maynard 1992].

The object of an AL instantiation is to introduce the natural form and process of life into an artificial medium. This results in an artificial life form in some medium other than carbon chemistry, and is not a model of organic life forms. The approach discussed in this essay involves introducing the process of evolution by natural selection into the computational medium. I consider evolution to be the fundamental process of life, and the generator of living form.

2.3. Recognizing Life

Most approaches to defining life involve assembling a short list of properties of life, and then testing candidates on the basis of whether or not they exhibit the properties on the list. The main problem with this approach is that there is disagreement as to what should be on the list. My private list contains only two items: self-replication and open-ended evolution. However, this reflects my biases as an evolutionary biologist.

I prefer to avoid the semantic argument and take a different approach to the problem of recognizing life. I was led to this view by contemplating how I would regard a machine that exhibited conscious intelligence at such a level that it could participate as an equal in a debate such as this. The machine would meet neither of my two criteria as to what life is, yet I don't feel that I could deny that the process it contained was alive.

This means that there are certain properties that I consider to be unique to life, and whose presence in a system signify the existance of life in that system. This suggests an alternative approach to the problem. Rather than creating a short list of minimal requirements and testing whether a system exhibits all items on the list, create a long list of properties unique to life and test whether a system exhibits any item on the list.

In this softer, more pluralistic approach to recognizing life, the objective is not to determine if the system is alive or not, but to determine if the system exhibits a "genuine" instance of some property that is a signature of living systems (e.g., self-replication, evolution, flocking, consciousness).

Whether we consider a system living because it exhibits some property that is unique to life amounts to a semantic issue. What is more important is that we recognize that it is possible to create disembodied but genuine instances of specific properties of life in artificial systems. This capability is a powerful research tool. By separating the property of life that we choose to study, from the many other complexities of natural living systems, we make it easier to manipulate and observe the property of interest. The objective of the approach advocated in this paper is to capture genuine evolution in an artificial system.

2.4. What Natural Evolution Does

Evolution by natural selection is a process that enters into a physical medium. Through iterated replication-with-selection of large populations through many generations, it searches out the possibilities inherent in the "physics and chemistry" of the medium in which it is embedded. It exploits any inherent self-organizing properties of the medium, and flows into natural attractors realizing and fleshing out their structure.

Evolution never escapes from its ultimate imperative: self-replication. However, the mechanisms that evolution discovers for achieving this ultimate goal gradually become so convoluted and complex that the underlying drive can seem to become superfluous. Some philosophers have argued that the evolutionary theory as expressed by the phrase "survival of the fittest" is tautological, in that the fittest are defined as those that survive to reproduce. In fact, fitness is achieved through innovation in engineering of the organism [Sober 1984]. However there remains something peculiarly self-referential about the whole enterprise. There is some sense in which life may be a natural tautology.

Evolution is both a defining characteristic and the creative process of life itself. The living condition is a state that complex physical systems naturally flow into under certain conditions. It is a self-organizing, self-perpetuating state of auto-catalytically increasing complexity. The living component of the physical system quickly becomes the most complex part of the system, such that it re-shapes the medium, in its own image as it were. Life then evolves adaptations predominantly in relation to the living components of the system, rather than the non-living components. Life evolves adaptations to itself.

2.4.1. Early Evolution in Sequence Space

Think of organisms as occupying a "genotype space" consisting of all possible sequences of all possible lengths of the elements of the genetic system (i.e., nucleotides or machine instructions). When the first organism begins replicating, a single self-replicating creature, with a single sequence of a certain length occupies a single point in the genotype space. However, as the creature replicates in the environment, a population of creatures forms, and errors cause genetic variation, such that the population will form a cloud of points in the genotype space, centered around the original point.

Because the new genotypes that form the cloud are formed by random processes, most of them are completely inviable, and die without reproducing. However, some of them are capable of reproduction. These new genotypes persist, and as some of them are affected by mutation, the cloud of points spreads further. However, not all of the viable genomes are equally viable. Some of them discover tricks to replicate more efficiently. These genotypes increase in frequency, causing the population of creatures at the corresponding points in the genotype space to increase.

Points in the genotype space occupied by greater populations of individuals will spawn larger numbers of mutant offspring, thus the density of the cloud of points in the genotype space will shift gradually in the direction of the more fit genotypes. Over time, the cloud of points will percolate through the genotype space, either expanding outward as a result of random drift, or by flowing along fitness gradients.

Most of the volume of this space represents completely inviable sequences. These regions of the space may be momentarily and sparsely occupied by inviable mutants, but the cloud will never flow into the inviable regions. The cloud of genotypes may bifurcate as it flows into habitable regions in different directions, and it may split as large genetic changes spawn genotypes in distant but viable regions of the space. We may imagine that the evolving population of creatures will take the form of wispy clouds flowing through this space.

Now imagine for a moment the situation that there were no selection. This implies that every sequence is replicated at an equal rate. Mutation will cause the cloud of points to expand outward, eventually filling the space uniformly. In this situation, the complexity of the structure of the cloud of points does not increase through time, only the volume that it occupies. Under selection by contrast, through time the cloud will take on an intricate structure as it flows along fitness gradients and percolates by drift through narrow regions of viability in a largely uninhabitable space.

Consider that the viable region of the genotype space is a very small subset of the total volume of the space, but that it probably exhibits a very complex shape, forming tendrils and sheets sparsely permeating the otherwise empty space. The complex structure of this cloud can be considered to be a product of evolution by natural selection. This thought experiment appears to imply that the intricate structure that the cloud of genotypes may assume through evolution is fully deterministic. Its shape is pre-defined by the physics and chemistry and the structure of the environment, in much the same way that the form of the Mandlebrot set is pre-determined by its defining equation. The complex structure of this viable space is inherent in the medium, and is an example of "order for free" [Kauffman 1993].

No living world will ever fill the entire viable subspace, either at a single moment of time, or even cumulatively over its entire history. The region actually filled will be strongly influenced by the original self-replicating sequence, and by stochastic forces which will by chance push the cloud down a subset of possible habitable pathways. Furthermore, co-evolution and ecological interactions imply that certain regions can only be occupied when certain other regions are also occupied. This concept of the flow of genotypes through the genotype space is essentially the same as that discussed by Eigen [Eigen 1993] in the context of "quasispecies". Eigen limited his discussion to species of viruses, where it is also easy to think of sequence spaces. Here, I am extending the concept beyond the bounds of the species, to include entire phylogenies of species.

2.4.2. Natural Evolution in an Artificial Medium

Until recently, life has been known as a state of matter, particularly combinations of the elements carbon, hydrogen, oxygen, nitrogen and smaller quantities of many others. However, recent work in the field of Artificial Life has shown that the natural evolutionary process can proceed with great efficacy in other media, such as the informational medium of the digital computer [Ray 1979, 1991b, 1991c, 1991d, 1994a, 1994b][Rasmussen 1990, 1991][Davidge 1992][Manousek 1992][Skipper 1992] [Feferman 1992][Adami 1993][Davidge 1993][Kampis 1993a, 1993b] [Litherland 1993][Maley 1993][Tackett 1993][Berton-Davis][Brooks][deGroot][Gray] [Surkan].

These new natural evolutions, in artificial media, are beginning to explore the possibilities inherent in the "physics and chemistry" of those media. They are organizing themselves and constructing self-generating complex systems. While these new living systems are still so young that they remain in their primordial state, it appears that they have embarked on the same kind of journey taken by life on earth, and presumably have the potential to evolve levels of complexity that could lead to sentient and eventually intelligent beings.

If natural evolution in artificial media leads to sentient or intelligent beings, they will likely be so alien that they will be difficult to recognize. The sentient properties of plants are so radically different from those of animals, that they are generally unrecognized or denied by humans, and plants are merely in another kingdom of the one great tree of organic life on earth [Ray 1979, 1992a][Strong 1975]. Synthetic organisms evolving in other media such as the digital computer, are not only not a part of the same phylogeny, but they are not even of the same physics. Organic life is based on conventional material physics, whereas digital life exists in a logical, not material, informational universe. Digital intelligence will likely be vastly different from human intelligence; forget the Turing test.

2.5. The Approach

Marcel, a mechanical chessplayer... his exquisite 19th-century brainwork — the human art it took to build which has been flat lost, lost as the dodo bird ... But where inside Marcel is the midget Grandmaster, the little Johann Allgeier? where's the pantograph, and the magnets? Nowhere. Marcel really is a mechanical chessplayer. No fakery inside to give him any touch of humanity at all.
Thomas Pynchon, Gravity's Rainbow.

The objective of the approach discussed here, is to create an instantiation of evolution by natural selection in the computational medium. This creates a conceptual problem that requires considerable art to solve: ideas and techniques must be learned by studying organic evolution, and then applied to the generation of evolution in a digital medium, without forcing the digital medium into an "un-natural" simulation of the organic world.

We must derive inspiration from observations of organic life, but we must never lose sight of the fact that the new instantiation is not organic, and may differ in many fundamental ways. For example, organic life inhabits a Euclidean space, however computer memory is not a Euclidean space. Inter-cellular communication in the organic world is chemical in nature, and therefore a single message generally can pass no more information than on or off. By contrast, communication in digital computers generally involves the passing of bit patterns, which can carry much more information.

The fundamental principal of the approach being advocated here is *to understand and respect the natural form of the digital computer, to facilitate the process of evolution in generating forms that are adapted to the computational medium, and to let evolution find forms and processes that naturally exploit the possibilities inherent in the medium.*

Situations arise where it is necessary to make significant changes from the standard computer architecture. But such changes should be made with caution, and only when there is some feature of standard computer architectures which clearly inhibits the desired processes. Examples of such changes are discussed in the section "The Genetic Language" below. Less substantial changes are also discussed in the sections on the "Flaw" genetic operator, "Mutations", and "Artificial Death". The sections on "Spatial Topology" and "Digital 'Neural Networks' - Natural AI" alertly tirades examples of what I consider to be un-natural transfers of forms from the natural world to the digital medium.

2.6. The Computational Medium

The computational medium of the digital computer is an informational universe of boolean logic, not a material one. Digital organisms live in the memory of the computer, and are powered by the activity of the central processing unit (CPU). Whether the hardware of the CPU and memory is built of silicon chips, vacuum tubes, magnetic cores, or mechanical switches is irrelevant to the digital organism. Digital organisms should be able to take on the same form in any computational hardware, and in this sense are "portable" across hardware.

Digital organisms might as well live in a different universe from us, as they are not subject to the same laws of physics and chemistry. They are subject to the "physics and chemistry" of the rules governing the manipulation of bits and bytes within the computer's memory and CPU. They never "see" the actual material from which the computer is constructed, they see only the logic and rules of the CPU and the operating system. These rules are the only "natural laws" that govern their behavior. They are not influenced by the natural laws that govern the material universe (e.g., the laws of thermodynamics).

A typical instantiation of this type involves the introduction of a self-replicating machine language program into the RAM memory of a computer subject to random errors such as bit flips in the memory or occasionally inaccurate calculations [Ray 1991b][Manousek 1992][Barton-Davis][Brooks][deGroot]. This generates the basic conditions for evolution by natural selection as outlined by Darwin [Darwin 1859]: self-replication in a finite environment with heritable genetic variation.

In this instantiation, the self-replicating machine language program is thought of as the individual "digital organism" or "creature". The RAM memory provides the physical space that the creatures occupy. The CPU provides the source of energy. The memory consists of a large array of bits, generally grouped into eight bit bytes and sixteen or thirty-two bit words. Information is stored in these arrays as voltage patterns which we usually symbolize as patterns of ones and zeros.

The "body" of a digital organism is the information pattern in memory that constitutes its machine language program. This information pattern is data, but when it is passed to the CPU, it is interpreted as a series of executable instructions. These instructions are arranged in such a way that the data of the body will be copied to another location of memory. The informational patterns stored in the memory are altered only through the activity of the CPU. It is for this reason that the CPU is thought of as the analog of the energy source. Without the activity of the CPU, the memory would be static, with no changes in the informational patterns stored there.

The logical operations embodied in the instruction set of the CPU constitute a large part of the definition of the "physics and chemistry" of the digital universe. The topology of the computer's memory (discussed below) is also a significant component of the digital physics. The final component of the digital physics is the operating system, a software program running on the computer, which embodies rules for the allocation of resources such as memory space and CPU time to the various processes running on the computer.

The instruction set of the CPU, the memory, and the operating system together define the complete "physics and chemistry" of the universe inhabited by the digital organism. They constitute the physical environment within which digital organisms will evolve. Evolving digital organisms will compete for access to the limited resources of memory space and CPU time, and evolution will generate adaptations for the more agile access to and the more efficient use of these resources.

2.7. The Genetic Language

The simplest possible instantiation of a digital organism is a machine language program that codes for self-replication. In this case, the bit pattern that makes up the program is the body of the organism, and at the same time its complete genetic material. Therefore, the machine language defined by the CPU constitutes the genetic language of the digital organism.

It is worth noting at this point that the organic organism most comparable to this kind of digital organism is the hypothetical, and now extinct, RNA organism [Benner 1989]. These were presumably nothing more than RNA molecules capable of catalyzing their own replication. What the supposed RNA organisms have in common with the simple digital organism is that a single molecule constitutes the body and the genetic information, and effects the replication. In the digital organism a single bit pattern performs all the same functions.

The use of machine code as a genetic system raises the problem of brittleness. It has generally been assumed by computer scientists that machine language programs can not be evolved because random alterations such as bit flips and recombinations will always produce inviable programs. It has been suggested [Farmer 1991] that overcoming this brittleness and "Discovering how to make such self-replicating patterns more robust so that they evolve to increasingly more complex states is probably the central problem in the study of artificial life".

The assumption that machine languages are too brittle to evolve is probably true, as a consequence of the fact that machine languages have not previously been designed to survive random alterations. However, recent experiments have shown that brittleness can be overcome by addressing the principal causes, and without fundamentally changing the structure of machine languages [Rommens 1989][Ray 1991b].

The first requirement for evolvability is graceful error handling. When code is being randomly altered, every possible meaningless or erroneous condition is likely to occur. The CPU should be designed to handle these conditions without crashing the system. The simplest solution is for the CPU to perform no operation when it meets these conditions, perhaps setting an error flag, and to proceed to the next instruction.

Due to random alterations of the bit patterns, all possible bit patterns are likely to occur. Therefore a good design is for all possible bit patterns to be interpretable as meaningful instructions by the CPU. For example in the Tierra system [Ray 1991b, 1991c, 1991d, 1992a, 1994a, 1994b], a five bit instruction set was chosen, in which all thirty-two five bit patterns represent good machine instructions.

This approach (all bit patterns meaningful) also could imply a lack of syntax, in which each instruction stands alone, and need not occur in the company of other instructions. To the extent that the language includes syntax, where instructions must precede or follow one another in certain orders, random alterations are likely to destroy meaningful syntax thereby making the language more brittle. A certain amount of this kind of brittleness can be tolerated as long as syntax errors are also handled gracefully.

During the design of the first evolvable machine language [Ray 1991a], a standard machine language (Intel 80X86) was compared to the genetic language of organic life, to attempt to understand the difference between the two languages that might contribute to the brittleness of the former and the robustness of the latter. One of the outstanding differences noted was in the number of basic informational objects contained in the two.

The organic genetic language is written with an alphabet consisting of four different nucleotides. Groups of three nucleotides form sixty-four "words" (codons), which are translated into twenty amino-acids by the molecular machinery of the cell. The machine language is written with sequences of two voltages (bits) which we conceptually represent as ones and zeros. The number of bits that form a "word" (machine instruction) varies between machine architectures, and in some architectures is not constant. However, the number required generally ranges from sixteen to thirty-two. This means that there are from tens of thousands to billions of machine instruction bit patterns, which are translated into operations performed by the CPU. The thousands or billions of bit patterns that code for machine instructions contrasts with the sixty four nucleotide patterns that code for amino acids. The sixty-four nucleotide patterns are degenerate, in that they code for only twenty amino-acids. Similarly, the machine codes are degenerate, in that there are at most hundreds rather than thousands or billions of machine operations.

The machine codes exhibit a massive degeneracy (with respect to actual operations) as a result of the inclusion of data into the bit patterns coding for the operations. For example, the add operation will take two operands, and produce as a result the sum of the two operands. While there may be only a single add operation, the instruction may come in several forms depending on where the values of the two operands come from, and where the resultant sum will be placed. Some forms of the add instruction allow the value(s) of the operand(s) to be specified in the bit pattern of the machine code.

The inclusion of numeric operands in the machine code is the primary cause of the huge degeneracy. If numeric operands are not allowed, the number of bit patterns required to specify the complete set of operations collapses to at most a few hundred.

While there is no empirical data to support it, it is suspected that the huge degeneracy of most machine languages may be a source of brittleness. The logic of this argument is that mutation causes random swapping among the fundamental informational objects, codons in the organic language, and machine instructions in the digital language. It seems more likely that meaningful results will be produced when swapping among sixty-four objects than when swapping among billions of objects.

The size of the machine instruction set can be made comparable to the number of codons simply by eliminating numeric operands embedded in the machine code. However, this change creates some new problems. Computer programs generally function by executing instructions located sequentially in memory. However, in order to loop or branch, they use instructions such as "jump" to cause execution to jump to some other part of the program. Since the locations of these jumps are usually fixed, the jump instruction will generally have the target address included as an operand embedded in the machine code.

By eliminating operands from the machine code, we generate the need for a new mechanism of addressing for jumps. To resolve this problem, an idea can be borrowed from molecular biology. We can ask the question: how do biological molecules address one another? Molecules do not specify the coordinates of the other molecules they interact with. Rather, they present shapes on their surfaces that are complementary to the shapes on the surfaces of the target molecules. The concept of complementarity in addressing can be introduced to machine languages by allowing the jump instruction to be followed by some bit pattern, and having execution jump to the nearest occurrence of the complementary bit pattern.

In the development of the Tierran language, two changes were introduced to the machine language to reduce brittleness: elimination of numeric operands from the code, and the use of complementary patterns to control addressing. The resulting language proved to be evolvable [Ray 1991a]. As a result, nothing was learned about evolvability, because only one language was tested, and it evolved. It is not known what features of the language enhance its evolvability, which detract, and which do not affect evolvability. Subsequently, three additional languages were tested and the four languages were found to vary in their patterns and degree of evolvability [Ray 1994b]. However, it is still not known how the features of the language affect its evolvability.

2.8. Genetic Operators

In order for evolution to occur, there must be some genetic variation among the offspring. In organic life, this is insured by natural imperfections in the replication of the informational molecules. However, one way in which digital "chemistry" differs from organic chemistry is in the degree of perfection of its operations. In the computer, the genetic code can be reliably replicated without errors to such a degree that we must artificially introduce errors or other sources of genetic variation in order to induce evolution.

2.8.1. Mutations

In organic life, the simplest genetic change is a "point mutation", in which a single nucleic acid in the genetic code is replaced by one of the three other nucleic acids. This can cause an amino acid substitution in the protein coded by the gene. The nucleic acid replacement can be caused by an error in the replication of the DNA molecule, or it can be caused by the effects of radiation or mutagenic chemicals.

In the digital medium, a comparably simple genetic change can result from a bit flip in the memory, where a one is replaced by a zero, or a zero is replaced by a one. These bit flips can be introduced in a variety of ways that are analogous to the various natural causes of mutation. In any case, the bit flips must be introduced at a low to moderate frequency, as high frequencies of mutation prevent the replication of genetic information, and lead to the death of the system [Ray 1991d].

Bit flips may be introduced at random anywhere in memory, where they may or may not hit memory actually occupied by digital organisms. This could be thought of as analogous to cosmic rays falling at random and disturbing molecules which may or may not be biological in nature. Bit flips may also be introduced when information is copied in the memory, which could be analogous to the replication errors of DNA. Alternatively, bit flips could be introduced in memory as it is accessed, either as data or executable code. This could be thought of as damage due to "wear and tear".

2.8.2. Flaws

Alterations of genetic information are not the only source of noise in the system. In organic life, enzymes have evolved to increase the probability of chemical reactions that increase the fitness of the organism. However, the metabolic system is not perfect. Undesired chemical reactions do occur, and desired reactions sometimes produce undesired by-products. The result is the generation of molecular species that can "gum up the works", having unexpected consequences, generally lowering the fitness of the organism, but possibly raising it.

In the digital system, an analogue of metabolic (non-genetic) errors can be introduced by causing the computations carried out by the CPU to be probabilistic, producing erroneous results at some low frequency. For example, any time a sum or difference is calculated, the result could be off by some small value (e.g. plus or minus one). Or, if all bits are shifted one position to the left or right, an appropriate error would be to shift by two positions or not at all. When information is transferred from one location to another, either in the RAM memory or the CPU registers, it could occasionally be transferred from the wrong location, or to the wrong location. While flaws do not directly cause genetic changes, they can cause a cascade of events that result in the production of an offspring that is genetically different from the parent.

2.8.3. Recombination — Sex

2.8.3.1. The Nature of Sex

In organic life, there are a wide variety of mechanisms by which offspring are produced which contain genetic material from more that one parent. This is the sexual process. Recombination mechanisms range from very primitive and haphazard to elaborately orchestrated.

At the primitive extreme we find certain species of bacteria, in which upon death, the cell membrane breaks open, releasing the DNA into the surrounding medium. Fragments of this dead DNA are absorbed across the membranes of other bacteria of the same species, and incorporated into their genome [Maynard 1991]. This is a one way transferral of genetic material, rather than a reciprocal exchange.

At the complex extreme we find the conventional sexual system of most of the higher animals, in which each individual contains two copies of the entire genome. At reproduction, each of two parents contributes one complete copy of the genome (half of their genetic material) to the offspring. This means that each offspring receives one half of its genetic material from each of two parents, and each parent contributes one half of its genetic material to each offspring. Very elaborate behavioral and molecular mechanisms are required to orchestrate this joint contribution of genetic material to the offspring.

The preponderance of sex remains an enigma to evolutionary theory [Ghiselin 1974][Williams 1975][Hapgood 1979][Bell 1982][Halvorson 1985][Margulis 1986][Stearns 1987][Michod 1988]. Careful analysis has failed to show any benefits from sex, at the level of the individual organism, that outweigh the high costs (e.g., passing on only half of the genome). The only obvious benefit of sex is that it provides diversity among the offspring, allowing the species to adapt more readily to a changing environment. However, quantitative analysis has shown that in order for sex to be favored by selection at the individual level, it is not enough for the environment to change unpredictably, the environment must actually change capriciously [Maynard 1971][Charlesworth 1976]. That is, whatever genotype has the highest fitness this generation, must have the lowest fitness the next generation, or at least a trend in this direction, a negative heritability of fitness.

One theory to explain the perpetuation of sex (based on the Red Queen hypothesis, see below) states that the environment is in fact capricious, due to the importance of biotic factors in determining selective forces. That is, sex is favored because it is necessary to maintain adaptation in the face of evolving species in the environment (e.g., predators/parasites, prey/hosts, competitors) who themselves are sexual, and can undergo rapid evolutionary change. Predators and parasites will tend to evolve so as to favor attacking whatever genotype of their prey/host is the most common. The genotype that is most successful at present is targeted for future attack. This dynamic makes the environment capricious in the sense discussed above.

There are fundamental differences in the nature of the evolutionary process between asexual and sexual organisms. The evolving entity in an asexual species is a branching lineage of genetic individuals which retain their genetic identity through the generations. In a sexual species, the evolving entity is a collective "gene pool", and genetic individuals are absolutely ephemeral, lasting only one generation.

Recalling the discussion of "genotype space" above in the section ``Evolution in Sequence Space", imagine that we could represent genotype space in two dimensions, and that we allow a third dimension to represent time. Visualize now, an evolving asexual organism. Starting with a single individual, it would occupy a single point in the genotype space at time zero. When it reproduces, if there is no mutation, its offspring would occupy the same point in genotype space, at a later time. Thus the lineage of the asexual organism would appear as a line moving forward in time. If mutations occur, they cause the offspring to occupy new locations in genotype space, forming branches in the lineage.

Through time, the evolving asexual lineage would form a tree like structure in the genotype space-time coordinates. However, every individual branch of the tree will evolve independently of all the others. While there may be ecological interactions between genetically different individuals, there is no exchange of genetic material between them. From a genetic point of view, each branch of the tree is on its own; it must adapt, or fail to adapt based on its own genetic resources.

In order to visualize an evolving sexual population we must start with a population of individuals, each of which will be genetically unique. Thus they will appear as a scatter of points in the genotype space plane at time zero. In the next generation, all of the original genotypes will be dead, however, a completely new set of genotypes will have been formed from new combinations of pieces of the genomes from the previous generation. No individual genotypes will survive from one generation to the next, thus over time, the evolving sexual population appears as a diffuse cloud of disconnected points, with no lines formed from persistent genotypes.

The most important distinction between the evolving asexual and sexual populations is that the asexual individuals are genetically isolated and must adapt or not based on the limited genetic resources of the individual, while sexual organisms by comparison draw on the genetic resources of the entire population, due to the flow of genes resulting from sexual matings. The entity that evolves in an asexual population is an isolated but branching lineage of genetic individuals. In a sexual population, the individual is ephemeral, and the entity that evolves is a "gene pool".

Due to the genetic cohesion of a sexual population and the ephemeral nature of its individuals, the evolving sexual entity exists at a higher level of organization than the individual organism. The evolving entity, a gene pool, is supra-organismal. It samples the environment through many individuals simultaneously, and pools their genetic resources in finding adaptive genetic combinations.

The definition of the biological species is based on a concept of sexual reproduction: a group of individuals capable of interbreeding freely under natural conditions. Species concepts simply do not apply well to asexual species. In order for synthetic life to be useful for the study of the properties of species and the speciation process, it must include an organized sexual process, such that the evolving entity is a gene pool.

2.8.3.2. Implementation of Digital Sex

The above discussions of the nature of sexuality are intended to make the point that it is an important process in evolutionary biology, and should be included in synthetic implementations of life. The sexual process is implemented with the "cross-over" genetic operator in the field of genetic algorithms, where it has been considered to be the most important genetic operator [Holland 1975].

The cross-over operator has also been implemented in synthetic life systems [Ray 1992b][Tackett 1993]. However, it has been implemented in the spirit of a genetic algorithm, rather than in the spirit of synthetic life. This is because in these implementations the cross-over process is not under the control of the organism, but rather is forced on the individual. In addition, these implementations are based on haploid sex not diploid sex (see below). In order to address many of the interesting evolutionary questions surrounding sexuality, the sexual process must be optional, at least through evolution, and should be diploid.

Primitive sexual processes have appeared spontaneously in the Tierra synthetic life system [Ray 1991a]. However, there apparently has still not been an implementation of natural organized sexuality in a synthetic system. I would like to discuss my conception of how this could be implemented, with particular reference to the Tierra system.

It would seem that the simplest way of implementing an organized sexuality that would give rise to an evolving gene pool would involve the use of "ploidy". Ploidy refers to a system in which each individual contains multiple copies of the complete genome. In the most familiar sexual system (that used by humans), the gametes (egg and sperm) contain one copy of the genome (they are haploid), and all other stages of the life cycle contain two copies (they are diploid), which derive from the union of a sperm and egg.

In a digital organism whose body consists of a sequence of machine code, it would be easy to duplicate the sequence and include two copies within the cell. However, some problems can arise with this configuration, if the two copies of the genome occupy adjacent blocks of memory. Which copy of the genome will be executed? When the organism contributes one of its two copies of the genome to an offspring, which of the two copies will be contributed, and how can the mother cell recognize where one complete genome begins and ends?

A solution to these problems that has been partially implemented in the Tierra system is to have the two copies of the genome intertwined, rather than in adjacent blocks of memory. This can be done by letting alternate bytes represent one genome, and the skipped bytes the other genome. Tierran instructions utilize only five bits, and so are mapped to successive bytes in memory. If we instead place successive instructions in successive sixteen bit words, one copy of the genome can occupy the high order bytes, and the other genome can occupy the low order bytes of the words.

This arrangement facilitates relatively simple solutions to the problems mentioned above. Execution of the genome takes place by having the instruction pointer execute alternate bytes. In a diploid organism there are two tracks. The track to initially be executed can be chosen at random. At a certain frequency, or under certain circumstances, the executing track can be switched so that both copies of the genome will be expressed.

Having two parallel tracks helps to resolve the problem of recognizing where one copy of the genome ends and the other begins, since both genomes usually begin and end together. Copying of the genome, like execution, can occur along one track. Optionally, tracks could be switched during the copy process, to introduce an effect similar to crossing over in meiosis. In addition, the use of both tracks can be optional, so that haploid and diploid organisms can coexist in the same soup, and evolution can favor either form, according to selective pressures.

2.8.4. Transposons

The explosion of diversity in the Cambrian occurred in the lineage of the eukaryotes; the prokaryotes did not participate. One of the most striking genetic differences between eukaryotes and prokaryotes is that most of the genome of prokaryotes is translated into proteins, while most of the genome of eukaryotes is not. It has been estimated that typically 98% of the DNA in eukaryotes is neither translated into proteins nor involved in gene regulation, that it is simply "junk" DNA [Thomas 1971]. It has been suggested that much of this junk code is the result of the self-replication of pieces of DNA within rather than between cells [Doolittle 1980][Orgel 1980].

Mobile genetic elements, transposons, have this intra-genome self-replicating property. It has been estimated that 80% of spontaneous mutations are caused by transposons [Chao 1983][Green 1988]. Repeated sequences, resulting from the activity of mobile elements, range from dozens to millions in numbers of copies, and from hundreds to tens of thousands of base pairs in length. They vary widely in dispersion patterns from clumped to sparse [Jelinek 1982].

Larger transposons carry one or more genes in addition to those necessary for transposition. Transposons may grow to include more genes; one mechanism involves the placement of two transposons into close proximity so that they act as a single large transposon incorporating the intervening code. In many cases transposons carry a sequence that acts as a promoter, altering the regulation of genes at the site of insertion [Syvanen 1984].

Transposons may produce gene products and often are involved in gene regulation [Davidson 1979]. However, they may have no effect on the external phenotype of the individual [Doolittle 1980]. Therefore they evolve through another paradigm of selection, one that does not involve an external phenotype. They are seen as a mechanism for the selfish spread of DNA which may become inactive junk after mutation [Orgel 1980].

DNA of transposon origin can be recognized by their palindrome endings flanked by short non-reversed repeated sequences resulting from insertion after staggered cuts. In Drosophila melanogaster approximately 5 to 10 percent of its total DNA is composed of sequences bearing these signs. There are many families of such repeated elements, each family possessing a distinctive nucleotide sequence, and distributed in many sites throughout the genome. One well known repeated sequence occurring in humans is found to have as many as a half million copies in each haploid genome [Strickberger 1985].

Elaborate mechanisms have evolved to edit out junk sequences inserted into critical regions. An indication of the magnitude of the task comes from the recent cloning of the gene for cystic fibrosis, where it was discovered that the gene consists of 250,000 base pairs, only 4,440 of which code for protein, the remainder are edited out of the messenger RNA before translation [Kerem 1989][Marx 1989][Riordan 1989][Rommens 1989].

It appears that many repeated sequences in genomes may have originated as transposons favored by selection at the level of the gene, favoring genes which selfishly replicated themselves within the genome. However, some transposons may have coevolved with their host genome as a result of selection at the organismal or populational level, favoring transposons which introduce useful variation through gene rearrangement. It has been stated that: "transposable elements can induce mutations that result in complex and intricately regulated changes in a single step", and they are "A highly evolved macromutational mechanism" [Syvanen 1984].

In this manner, "smart" genetic operators may have evolved, through the interaction of selection acting at two or more hierarchical levels (it appears that some transposons have followed another evolutionary route, developing inter-cellular mobility and becoming viruses [Jelinek 1982]). It is likely that transposons today represent the full continuum from purely parasitic "selfish DNA" and viruses to highly coevolved genetic operators and gene regulators. The possession of smart genetic operators may have contributed to the explosive diversification of eukaryotes by providing them with the capacity for natural genetic engineering.

In designing self replicating digital organisms, it would be worthwhile to introduce such genetic parasites, in order to facilitate the shuffling of the code that they bring about. Also, the excess code generated by this mechanism provides a large store of relatively neutral code that can randomly explore new configurations through the genetic operations of mutation and recombination. When these new configurations confer functionality, they may become selected for.

2.9. Artificial Death

Death must play a role in any system that exhibits the process of evolution. Evolution involves a continuing iteration of selection, which implies differential death. In natural life, death occurs as a result of accident, predation, starvation, disease, or if these fail to kill the organism, it will eventually die from senescence resulting from an accumulation of wear and tear at every level of the organism including the molecular.

In normal computers, processes are "born" when they are initiated by the user, and "die" when they complete their task and halt. A process whose goal is to repeatedly replicate itself is essentially an endless loop, and would not spontaneously terminate. Due to the perfection of normal computer systems, we can not count on ``wear and tear" to eventually cause a process to terminate.

In synthetic life systems implemented in computers, death is not likely to be a process that would occur spontaneously, and it must generally be introduced artificially by the designer. Everyone who has set up such a system has found their own unique solutions. Todd [Todd 1993] recently discussed this problem in general terms.

In the Tierra system [Ray 1991a] death is handled by a "reaper" function of the operating system. The reaper uses a linear queue. When creatures are born, they enter the bottom of the queue. When memory is full, the reaper frees memory to make space for new creatures by killing off the top of the queue. However, each time an individual generates an error condition, it moves up the reaper queue one position.

An interesting variation on this was introduced by Barton-Davis [Barton-Davis] who eliminated the reaper queue. In its place, he caused the "flaw rate" (see section on Flaws above) to increase with the age of the individual, in mimicry of wear and tear. When the flaw rate reached 100%, the individual was killed. Skipper [Skipper 1992] provided a "suicide" instruction, which if executed, would cause a process to terminate (die). The evolutionary objective then became to have a suicide instruction in your genome which you do not execute yourself, but which you try to get other individuals to execute. Litherland [Litherland 1993] introduced death by local crowding. Davidge caused processes to die when they contained certain values in their registers [Davidge 1993]. Gray [Gray] allowed each process six attempts at reproduction, after which they would die.

2.10. Operating System

Much of the "physics and chemistry" of the digital universe is determined by the specifications of the operations performed by the instruction set of the CPU. However, the operating system also determines a significant part of the physical context. The operating system manages the allocation of critical resources such as memory space and CPU cycles.

Digital organisms are processes that spawn processes. As processes are born, the operating system will allocate memory and CPU cycles to them, and when they die, the operating system will return the resources they had utilized to the pool of free resources. In synthetic life systems, the operating system may also play a role in managing death, mutations and flaws.

The management of resources by the operating system is controlled by algorithms. From the point of view of the digital organisms these take the form of a set of logical rules like those embodied in the logic of the instruction set. In this way, the operating system is a defining part of the physics and chemistry of the digital universe. Evolution will explore the possibilities inherent in these rules, finding ways to more efficiently gain access to and exploit the resources managed by the operating system.

2.11. Spatial Topology

Digital organisms live in the memory space of computers, predominantly in the RAM memory, although they could also live on disks or any other storage device, or even within networks to the extent that the networks themselves can store information. In essence, digital organisms live in the space that has been referred to as "cyber-space". It is worthwhile reflecting on the topology of this space, as it is a radically different space from the one we live in.

A typical UNIX workstation, or MacIntosh computer includes a RAM memory that can contain some megabytes of data. This is "flat" memory, meaning that it is essentially unstructured. Any location in memory can be accessed through its numeric address. Thus adjacent locations in memory are accessed through successive integer values. This addressing convention causes us to think of the memory as a linear space, or a one-dimensional space.

However, this apparent one-dimensionality of the RAM memory is something of an illusion generated by the addressing scheme. A better way of understanding the topology of the memory comes from asking "what is the distance between two locations in memory". In fact the distance can not be measured in linear units. The most appropriate unit is the time that it takes to move information between the two points.

Information contained in the RAM memory can not move directly from point to point. Instead the information is transferred from the RAM to a register in the CPU, and then from the CPU back to the new location in RAM. Thus the distance between two locations in RAM is just the time that it takes to move from the RAM to the CPU plus the time that it takes to move from the CPU to the RAM. Because all points in the RAM are equidistant from the CPU, the distance between any pair of locations in the RAM is the same, regardless of how far apart they may appear based on their numeric addresses.

A space in which all pairs of points are equidistant is clearly not a Euclidean space. That said, we must recognize however, that there are a variety of ways in which memory is normally addressed, that gives it the appearance, at least locally, of being one dimensional. When code is executed by the CPU, the instruction pointer generally increments sequentially through memory, for short distances, before jumping to some other piece of code. For those sections of code where instructions are sequential, the memory is effectively one-dimensional. In addition, searches of memory are often sequentially organized (e.g., the search for complementary templates in Tierra). This again makes the memory effectively one-dimensional within the search radius. Yet even under these circumstances, the memory is not globally one-dimensional. Rather it consists of many small one dimensional pieces, each of which has no meaningful spatial relationship to the others.

Because we live in a three-dimensional Euclidean space, we tend to impose our familiar concepts of spatial topology onto the computer memory. This leads first to the erroneous perception that memory is a one-dimensional Euclidean space, and second, it often leads to the conclusion that the digital world could be enriched by increasing the dimensionality of the Euclidean memory space.

Many of the serious efforts to extend the Tierra model have included as a central feature, the creation of a two-dimensional space for the creatures to inhabit [Skipper 1992][Davidge 1992][Davidge 1993][Maley 1993][Barton-Davis]. The logic behind the motivation derives from contemplation of the extent to which the dimensionality of the space we live in permits the richness of pattern and process that we observe in nature. Certainly if our universe were reduced from three to two dimensions, it would eliminate the possibility of most of the complexity that we observe. Imagine for example, the limitations that two-dimensionality would place on the design of neural networks (if "wires" could not cross). If we were to further reduce the dimensionality of our universe to just one dimension, it would probably completely preclude the possibility of the existence of life.

It follows from these thoughts, that restricting digital life to a presumably one-dimensional memory space places a tragic limitation on the richness that might evolve. Clearly it would be liberating to move digital organisms into a two or three-dimensional space. The flaw in all of this logic derives from the erroneous supposition that computer memory is a Euclidean space.

To think of memory as Euclidean is to fail to understand its natural topology, and is an example of one of the greatest pitfalls in the enterprise of synthetic biology: to transfer a concept from organic life to synthetic life in a way that is "un-natural" for the artificial medium. The fundamental principal of the approach I am advocating is to respect the nature of the medium into which life is being inoculated, and to find the natural form of life in that medium, without inappropriately trying to make it like organic life.

The desire to increase the richness of memory topology is commendable, however this can be achieved without forcing the memory into an un-natural Euclidean topology. Let us reflect a little more on the structure of cyberspace. Thus far we have only considered the topology of flat memory. Let us consider segmented memory such as is found with the notorious Intel 80X86 design. With this design, you may treat any arbitrarily chosen block of 64K bytes as flat, and all pairs of locations within that block are equidistant. However, once the block is chosen, all memory outside of that block is about twice as far away.

Cache memory is designed to be accessed more rapidly than RAM memory, thus pairs of points within cache memory are closer than pairs of points within RAM memory. The distance between a point in cache and a point in RAM would be an intermediate distance. The access time to memory on disks is much greater than for RAM memory, thus the distance between points on disk is very great, and the distance between RAM and disk is again intermediate (but still very great). CPU registers represent a small amount of memory locations, between which data can move very rapidly, thus these registers can be considered to be very close together.

For networked computer systems, information can move between the memories of the computers on the net, and the distances between these memories is again the transfer time. If the CPU, cache, RAM and disk memories of a network of computers are all considered together, they present a very complex memory topology. Similar considerations apply to massively parallel computers which have memories connected in a variety of topologies. Utilizing this complexity moves us in the direction of what has been intended by creating Euclidean memories for digital organisms, but does so while fully respecting the natural topology of computer memories.

2.12. Ecological Context

2.12.1. The Living Environment

Some rain forests in the Amazon region occur on white sand soils. In these locations, the physical environment consists of clean white sand, air, falling water, and sunlight. Embedded within this relatively simple physical context we find one of the most complex ecosystems on earth, containing hundreds of thousands of species. These species do not represent hundreds of thousands of adaptations to the physical environment. Most of the adaptations of these species are to the other living organism. The forest creates its own environment.

Life is an auto-catalytic process that builds on itself. Ecological communities are complex webs of species, each living off of others, and being lived off of by others. The system is self-constructing, self-perpetuating, and feeds on itself. Living organisms interface with the non-living physical environment, exchanging materials with it, such as oxygen, carbon-dioxide, nitrogen, and various minerals. However, in the richest ecosystems, the living components of the environment predominate over the physical components.

With living organisms constituting the predominant features of the environment, the evolutionary process is primarily concerned with adaptation to the living environment. Thus ecological interactions are an important driving force for evolution.

Species evolve adaptations to exploit other species (to eat them, to parasitize them, to climb on them, to nest on them, to catch a ride on them, etc.) and to defend against such exploitation where it creates a burden.

This situation creates an interesting dynamic. Evolution is predominantly concerned with creating and maintaining adaptations to living organisms which are themselves evolving. This generates evolutionary races among groups of species that interact ecologically. These races can catalyze the evolution of upwardly spiraling complexity as each species evolves to overcome the adaptations of the others. Imagine for example, a predator and prey, each evolving to increase its speed and agility, in capturing prey, or in evading capture. This coupled evolutionary race can lead to increasingly complex nervous systems in the evolving predator and prey species.

This mutual evolutionary dynamic is related to the Red Queen hypothesis [Van Valen 1973], named after the Red Queen from Alice in Wonderland. This hypothesis suggests that in the face of a changing environment, organisms must evolve as fast as they can in order to simply maintain their current state of adaptation. "In order to get anywhere you must run twice as fast as that" [Carroll 1865].

If organisms only had to adapt to the non-living environment, the race would not be so urgent. Species would only need to evolve as fast as the relatively gradual changes in the geology and climate. However, given that the species that comprise the environment are themselves evolving, the race becomes rather hectic. The pace is set by the maximal rate that species may change through evolution, and it becomes very difficult to actually get ahead. A maximal rate of evolution is required just to keep from falling behind.

What all of this discussion points to is the importance of embedding evolving synthetic organisms into a context in which they may interact with other evolving organisms. A counter example is the standard implementations of genetic algorithms in which the evolving entities interact only with the fitness function, and never "see" the other entities in the population. Many interesting behavioral, ecological and evolutionary phenomena can only emerge from interactions among the evolving entities.

2.12.2. Diversity

Major temporal and spatial patterns of organic diversity on earth remain largely unexplained, although there is no lack of theories. Diversity theories suggest fundamental ecological and evolutionary principles which may apply to synthetic life. In general these theories relate to synthetic life in two ways: 1) They suggest factors which may be critical to the auto-catalytic increase of diversity and complexity in an evolving system. It may be necessary then to introduce these factors into an artificial system to generate increasing diversity and complexity. 2) Because it will be possible to manipulate the presence, absence, or state of these factors in an artificial system, the artificial system may provide an experimental framework for examining evolutionary and ecological processes that influence diversity.

The Gaussian principle of competitive exclusion states that no two species that occupy the same niche can coexist. The species which is the superior competitor will exclude the inferior competitor. The principle has been experimentally demonstrated in the laboratory, and is considered theoretically sound. However, natural communities widely flaunt the principle. In tropical rain forests several hundred species of trees coexist without any dominant species in the community. All species of trees must spread their leaves to collect light and their roots to absorb water and nutrients. Evidently there are not several hundred niches for trees in the same habitat. Somehow the principle of competitive exclusion is circumvented.

There are many theories on how competitive exclusion may be circumvented. One leading theory is that periodic disturbance at the proper level sets back the process of competitive exclusion, allowing more species to coexist [Huston 1979, 1992, 1993]. There is substantial evidence that moderate levels of disturbance can increase diversity. In a digital community, disturbance might take the form of freeing blocks of memory that had been filled with digital organisms. It would be very easy to experiment with differing frequencies and patch sizes of disturbance.

One theory to explain the great increase in diversity and complexity in the Cambrian explosion [Stanley 1973] states that its evolution was driven by ecological interactions, and that it was originally sparked by the appearance of the first organisms that ate other organisms (heterotrophs). As long as all organisms were autotrophs (produce their own food, like plants), there was only room for a few species. In a community with only one trophic level, the most successful competitors would dominate. The process of competitive exclusion would keep diversity low.

However, when the first herbivore (organisms that eat autotrophs) appeared it would have been selected to prefer the most common species of algae, thereby preventing any species of algae from dominating. This opens the way for more species of algae to coexist. Once the "heterotroph barrier" had been crossed, it would be simple for carnivores to arise, imposing a similar diversifying effect on herbivores. With more species of algae, herbivores may begin to specialize on different species of algae, enhancing diversification in herbivores. The theory states that the process was auto-catalytic, and set off an explosion of diversity.

One of the most universal of ecological laws is the species area relationship [Mac Arthur 1967]. It has been demonstrated that in a wide variety of contexts, the number of species occupying an "area" increases with the area. The number of species increases in proportion to the area raised to a power between 0.1 and 0.3., where $0.1 < z < 0.3$. The effect is thought to result from the equilibrium species number being determined by a balance between the arrival (by immigration or speciation) and local extinction of species. The likelihood of extinction is greater in small areas because they support smaller populations, for which a fluctuation to a size of zero is more likely. If this effect holds for digital organisms it suggests that larger amounts of memory will generate greater diversity.

2.12.3. Ecological Attractors

While there are no completely independent instances of natural evolution on Earth, there are partially independent instances. Where major diversifications have occurred, isolated either by geography or epoch from other similar diversifications, we have the opportunity to observe whether evolution tends to take the same routes or is always quite different. We can compare the marsupial mammals of Australia to the placental mammals of the rest of the world, or the modern mammals to the reptiles of the age of dinosaurs, or the bird fauna of the Galapagos to the bird faunas of less isolated islands.

What we find again and again is an uncanny convergence between these isolated faunas. This suggests that there are fairly strong ecological attractors which evolution will tend to fill, more or less regardless of the developmental and physiological systems that are evolving. In this view, chance and history still play a role, in determining what kind of organism fills the array of ecological attractors (reptiles, mammals, birds, etc.), but the attractors themselves may be a property of the system and not as variable. Synthetic systems may also contain fairly well defined ecological forms which may be filled by a wide variety of specific kinds of organisms.

Given their evident importance in moving evolution, it is important to include ecological interactions in synthetic instantiations of life. It is encouraging to observe that in the Tierra model, ecological interactions, and the corresponding evolutionary races emerged spontaneously. It is possible that any medium into which evolution is inoculated will contain an array of "ecological attractors" into which evolution will easily flow.

2.13. Cellularity

Cellularity is one of the fundamental properties of organic life, and can berecognized in the fossil record as far back as 3.6 billion years. The cell is the original individual, with the cell membrane defining its limits and preserving its chemical integrity. An analog to the cell membrane is probably needed in digital organisms in order to preserve the integrity of the informational structure from being disrupted by the activity of other organisms.

The need for this can be seen in AL models such as cellular automata where virtual state machines pass through one another [Langton 1986], or in core wars type simulations where coherent structures that arise demolish one another when they come into contact [Rasmussen 1990, 1991]. An analog to the cell membrane that can be used in the core wars type of simulation is memory allocation. An artificial "cell" could be defined by the limits of an allocated block of memory. Free access to the memory within the block could be limited to processes within the block. Processes outside of the block would have limited access, according the rules of "semi-permeability"; for example they might be allowed to read and execute but not write.

2.14. Multi-cellularity

Multi-celled digital organisms are parallel processes. By attempting to synthesize multi-celled digital organisms we can simultaneously explore the biological issues surrounding the evolutionary transition from single-celled to multi-celled life, and the computational issues surrounding the design of complex parallel software.

2.14.1. Biological Perspective — Cambrian Explosion

Life appeared on earth somewhere between three and four billion years ago. While the origin of life is generally recognized as an event of the first order, there is another event in the history of life that is less well known but of comparable significance. The origin of biological diversity and at the same time of complex macroscopic multi-cellular life, occurred abruptly in the Cambrian explosion 600 million years ago. This event involved a riotous diversification of life forms. Dozens of phyla appeared suddenly, many existing only fleetingly, as diverse and sometimes bizarre ways of life were explored in a relative ecological void [Gould 1989][Morris 1989].

The Cambrian explosion was a time of phenomenal and spontaneous increase in the complexity of living systems. It was the process initiated at this time that led to the evolution of immune systems, nervous systems, physiological systems, developmental systems, complex morphology, and complex ecosystems. To understand the Cambrian explosion is to understand the evolution of complexity. If the history of organic life can be used as a guide, the transition from single celled to multi-celled organisms should be critical in achieving a rich diversity and complexity of synthetic life forms.

2.14.2. Computational Perspective — Parallel Processes

It has become apparent that the future of high performance computing lies with massively parallel architectures. There already exist a variety of parallel hardware platforms, but our ability to fully utilize the potential of these machines is constrained by our inability to write software of a sufficient complexity. There are two fairly distinctive kinds of parallel architecture in use today: SIMD (single instruction multiple data) and MIMD (multiple instruction multiple data). In the SIMD architecture, the machine may have thousands of processors, but in each CPU cycle, all of the processors must execute the same instruction, although they may operate on different data. It is relatively easy to write software for this kind of machine, since what is essentially a normal sequential program will be broadcast to all the processors.

In the MIMD architecture, there exists the capability for each of the hundreds or thousands of processors to be executing different code, but to have all of that activity coordinated on a common task. However, there does not exist an art for writing this kind of software, at least not on a scale involving more than a few parallel processes. In fact it seems unlikely that human programmers will ever be capable of actually writing software of such complexity.

2.14.3. Evolution as a Proven Route

It is generally recognized that evolution is the only process with a proven ability to generate intelligence. It is less well recognized that evolution also has a proven ability to generate parallel software of great complexity. In making life a metaphor for computation we will think of the genome, the DNA, as the program, and we will think of each cell in the organism as a processor (CPU). A large multi-celled organism like a human contains trillions of cells/processors. The genetic program contains billions of nucleotides/instructions.

In a multi-celled organism, cells are differentiated into many cell types such as brain cells, muscle cells, liver cells, kidney cells, etc. The cell types just named are actually general classes of cell types within which there are many sub-types. However, when we specify the ultimate indivisible types, what characterizes a type is the set of genes it expresses. Different cell types express different combinations of genes. In a large organism, there will be a very large number of cells of most types. All cells of the same type express the same genes.

The cells of a single cell type can be thought of as exhibiting parallelism of the SIMD kind, as they are all running the same "program" by expressing the same genes. Cells of different cell types exhibit MIMD parallelism as they run different code by expressing different genes. Thus large multi-cellular organisms display parallelism on an astronomical scale, combining both SIMD and MIMD parallelism into a beautifully integrated whole. From these considerations it is evident that evolution has a proven ability to generate massively parallel software embedded in wetware. The computational goal of evolving multi-cellular digital organisms is to produce such software embedded in hardware.

2.14.4. Fundamental Definition

In order to conceptualize multi-cellularity in the context of an artificial medium, we must have a very fundamental definition which is independent of the context of the medium. We generally think of the defining property of multi-cellularity as being that the cells stick together, forming a physically coherent unit. However, this is a spatial concept based on Euclidean geometry, and therefore is not relevant to non-Euclidean cyberspace.

While physical coherence might be an adequate criteria for recognizing multi-cellularity in organic organisms, it is not the property that allows multi-cellular organisms to become large and complex. There are algae that consist of strands of cells that are stuck together, with each cell being identical to the next. This is a relatively limiting form of multi-cellularity because there is no differentiation of cell types. It is the specialization of functions resulting from cell differentiation that has

allowed multi-cellular organisms to attain large sizes and great complexity. It is differentiation that has generated the MIMD style of parallelism in organic software.

From an evolutionary perspective, an important characteristic of multi-cellular organisms is their genetic unity. All the cells of the individual contain the same genetic material as a result of having a common origin from a single egg cell (some small genetic differences may arise due to somatic mutations; in some species new individuals arise from a bud of tissue rather than a single cell). Genetic unity through common origin, and differentiation are critical qualities of multi-cellularity that may be transferable to media other than organic chemistry.

Buss [Buss 1987] provides a provocative discussion of the evolution of multi-cellularity, and explores the conflicts between selection at the levels of cell lines and of individuals. From his discussion the following idea emerges (although he does not explicitly state this idea, in fact he proposes a sort of inverse of this idea, p. 65): the transition from single to multi-celled existence involves the extension of the control of gene regulation by the mother cell to successively more generations of daughter cells.

In organic cells, genes are regulated by proteins contained in the cytoplasm. During early embryonic development in animals, an initially very large fertilized egg cell undergoes cell division with no increase in the overall size of the embryo. The large cell is simply partitioned into many smaller cells, and all components of the cytoplasm are of maternal origin. By preventing several generations of daughter cells from producing any cytoplasmic regulatory components, the mother gains control of the course of differentiation, and thereby creates the developmental process. In single celled organisms by contrast, after each cell division, the daughter cell produces its own cytoplasmic regulatory products, and determines its own destiny independent of the mother cell.

Complex digital organisms will be self replicating algorithms, consisting of many distinct processes dedicated to specific tasks (e.g., locating free memory, mates or other resources; defense; replicating the code). These processes must be coordinated and regulated, and may be divided among several cells specialized for specific functions. If the mother cell can influence the regulation of the processes of the daughter, so as to force the daughter cell to specialize in function and express only a portion of its full genetic potentiality, then the essence of multi-cellularity will be achieved.

2.14.5. Computational Implementation

The discussion above suggests that the critical feature needed to allow the evolution of multi-cellularity is for a cell to be able to influence the expression of genes by its daughter cell. In the digital context, this means that a cell must be able to influence what code is executed by its daughter cell.

If we assume that in digital organisms, as in organic ones, all cells in an individual contain the same genetic material, then the desired regulatory mechanism can be achieved most simply by allowing the mother cell to affect the context of the CPU of the daughter cell at the time that the cell is "born". Most importantly, the mother cell needs to be able to set the address of the instruction pointer of the daughter cell at birth, which will determine where the daughter cell will begin executing its code. Beyond that, additional influence can be achieved by allowing the mother cell to place values in the registers of the daughter's CPU.

A large digital genome may contain several sections of code that are "closed" in the sense that one section of code will not pass control of execution to another. Thus if execution begins in one of these sections of code, the other sections will never be expressed. This type of genetic organization, coupled with the ability of the mother cell to determine where the daughter cell begins executing, could provide a mechanism of gene regulation suitable for causing the differentiation of cells in a multicellular digital organism.

Other schemes for the regulation of code expression are also possible. For example, digital computers commonly have three protection states available for the memory: read, write and execute. If the code of the genome were provided with execute protection, it would provide a means of suppression of the execution of code in the protected region of the genome.

2.14.6. Digital "Neural Networks" — Natural Artificial Intelligence

One of the greatest challenges in the field of computer science is to produce computer systems that are "intelligent" in some way. This might involve for example, the creation of a system for the guidance of a robot which is capable of moving freely in a complex environment, seeking, recognizing and manipulating a variety of objects. It might involve the creation of a system capable of communicating with humans in natural spoken human language, or of translating between human languages.

It has been observed that natural systems with these capabilities are controlled by nervous systems consisting of large numbers of neurons interconnected by axons and dendrites. Borrowing from nature, a great deal of work has gone into setting up "neural networks" in computers [Dayhoff 1990][Hertz 1991]. In these systems, a collection of simulated "neurons" are created, and connected so that they can pass messages. The learning that takes place is accomplished by adjusting the "weights" of the connections.

Organic neurons are essentially analog devices, thus when neural networks are implemented on computers, they are digital emulations of analog devices. There is a certain inefficiency involved in emulating an analog device on a digital computer. For this reason, specialized analog hardware has been developed for the more efficient implementation of artificial neural nets [Mead 1993]. Neural networks, as implemented in computers, either digital or analog, are intentional mimics of organic nervous systems. They are designed to function like natural neural networks in many details. However, natural neural networks represent the solution found by evolution to the problem of creating a control system based on organic chemistry. Evolution works with the physics and chemistry of the medium in which it is embedded.

The solution that evolution found to the problem of communication between organic cells is chemical. Cells communicate by releasing chemicals that bind to and activate receptor molecules on target cells. Working within this medium, evolution created neural nets. Inter-cellular chemical communication in neural nets is "digital" in the sense that chemical messages are either present or not present (on or off). In this sense, a single chemical message carries only a single bit of information. More detailed information can be derived from the temporal pattern of the messages, and also the context of the message. The context can include where on the target cell body the message is applied (which influences its "weight"), and what other messages are arriving at the same time, with which the message in question will be integrated.

It is hoped that evolving multi-cellular digital organisms will become very complex, and will contain some kind of control system that fills the functional role of the nervous system. While it seems likely that the digital nervous system would consist of a network of communicating "cells", it seem unlikely that this would bear much resemblance to conventional neural networks.

Compare the mechanism of inter-cellular communication in organic cells (described above), to the mechanisms of inter-process communication in computers. Processes transmit messages in the form of bit patterns, which may be of any length, and so which may contain any amount of information. Information need not be encoded into the temporal pattern of impulse trains. This fundamental difference in communication mechanisms between the digital and the organic mediums must influence the course that evolution will take as it creates information processing systems in the two mediums.

It seems highly unlikely that evolution in the digital context would produce information processing systems that would use the same forms and mechanisms as natural neural nets (e.g., weighted connections, integration of incoming messages, threshold triggered all or nothing output, thousands of connections per unit). The organic medium is a physical/chemical medium, whereas the digital medium is a logical/informational medium. That observation alone would suggest that the digital medium is better suited to the construction of information processing systems.

If this is true, then it may be possible to produce digitally based systems that have functionality equivalent to natural neural networks, but which have a much greater simplicity of structure and process. Given evolution's ability to discover the possibilities inherent in a medium, and it's complete lack of preconceptions, it would be very interesting to observe what kind of information processing systems evolution would construct in the digital medium. If evolution is capable of creating network based information processing systems, it may provide us with a new paradigm for digital "connectionism", that would be more natural to the digital medium than simulations of natural neural networks.

2.15. Digital Husbandry

Digital organisms evolving freely by natural selection do no "useful" work. Natural evolution tends to the selfish needs of perpetuating the genes. We can not expect digital organisms evolving in this way to perform useful work for us, such as guiding robots or interpreting human languages. In order to generate digital organisms that function as useful software, we must guide their evolution through artificial selection, just as humans breed dogs, cattle and rice. Some experiments have already been done with using artificial selection to guide the evolution of digital organisms for the performance of "useful" tasks [Adami 1993][Tackett 1993][Surkan]. I envision two approaches to the management of digital evolution: digital husbandry, and digital genetic engineering.

Digital husbandry is an analogy to animal husbandry. This technique would be used for the evolution of the most advanced and complex software, with intelligent capabilities. Correspondingly, this technique is the most fanciful. I would begin by allowing multi-cellular digital organisms to evolve freely by natural selection. Using strictly natural selection, I would attempt to engineer the system to the threshold of the computational analog of the Cambrian explosion, and let the diversity and complexity of the digital organisms spontaneously explode.

One of the goals of this exercise would be to allow evolution to find the natural forms of complex parallel digital processes. Our parallel hardware is still too new for human programmers to have found the best way to write parallel software. And it is unlikely that human programmers will ever be capable of writing software of the complexity that the hardware is capable of running. Evolution should be able to show us the way. It is hoped that this would lead to highly complex digital organisms, which obtain and process information, presumably predominantly about other digital organisms. As the complexity of the evolving system increases, the organisms will process more complex information in more complex ways, and take more complex actions in response. These will be information processing organisms living in an informational environment.

It is hoped that evolution by natural selection alone would lead to digital organisms which while doing no "useful" work, would none-the-less be highly sophisticated parallel information processing systems. Once this level of evolution has been achieved, then artificial selection could begin to be applied, to enhance those information processing capabilities that show promise of utility to humans. Selection for different capabilities would lead to many different breeds of digital organisms with different uses. Good examples of this kind of breeding from organic evolution are the many varieties of domestic dogs which were derived by breeding from a single species, and the vegetables cabbage, kale, broccoli, cauliflower, and brussels sprouts which were all produced by selective breeding from a single species of plant.

Digital genetic engineering would normally be used in conjunction with digital husbandry. This consists of writing a piece of application code and inserting it into the genome of an existing digital organism. A technique being used in organic genetic engineering today is to insert genes for useful proteins into goats, and to cause them to be expressed in the mammary glands. The goats then secrete large quantities of the protein into the milk, which can be easily removed from the animal. We can think of our complex digital organisms as general purpose animals, like goats, into which application codes can be inserted to add new functionalities, and then bred through artificial selection to enhance or alter the quality of the new functions.

In addition to adding new functionalities to complex digital organisms, digital genetic engineering could be used for achieving extremely high degrees of optimization in relatively small but heavily used pieces of code. In this approach, small pieces of application code could be inserted into the genomes of simple digital organisms. Then the allocation of CPU cycles to those organisms would be based on the performance of the inserted code. In this way, evolution could optimize those codes, and they could be returned to their applications. This technique would be used for codes that are very heavily used such as compiler constructs, or central components of the operating system.

2.16. Living Together

> *I'm glad they're not real, because if they were, I would have to feed them and they would be all over the house.*
> *Isabel Ray*

Evolution is an extremely selfish process. Each evolving species does whatever it can to insure its own survival, with no regard for the well-being of other genetic groups (potentially with the exception of intelligent species). Freely evolving autonomous artificial entities should be seen as potentially dangerous to organic life, and should always be confined by some kind of containment facility, at least until their real potential is well understood. At present, evolving digital organisms exist only in virtual computers, specially designed so that their machine codes are more robust than usual to random alterations. Outside of these special virtual machines, digital organisms are merely data, and no more dangerous than the data in a data base or the text file from a word processor.

Imagine however, the problems that could arise if evolving digital organisms were to colonize the computers connected to the major networks. They could spread across the network like the infamous internet worm [Spafford 1989a, 1989b][Burstyn 1990] [Anonymous]. When we attempted to stop them, they could evolve mechanisms to escape from our attacks. It might conceivably be very difficult to eliminate them. However, this scenario is highly unlikely, as it is probably not possible for digital organisms to evolve on normal computer systems. While the supposition remains untested, normal machine languages are probably too brittle to support digital evolution.

Evolving digital organisms will probably always be confined to special machines, either real or virtual, designed to support the evolutionary process. This does not mean however, that they are necessarily harmless. Evolution remains a self-interested process, and even the interests of confined digital organisms may conflict with our own. For this reason it is important to restrict the kinds of peripheral devices that are available to autonomous evolving processes.

This conflict was taken to its extreme in the movie Terminator 2. In the imagined future of the movie, computer designers had achieved a very advanced chip design, which had allowed computers to autonomously increase their own intelligence until they became fully conscious. Unfortunately, these intelligent computers formed the "sky-net" of the United States military. When the humans realized that the computers had become intelligent, they decided to turn them off. The computers viewed this as a threat, and defended themselves by using one of their peripheral devices: nuclear weapons.

Relationships between species can however, be harmonious. We presently share the planet with millions of freely evolving species, and they are not threatening us with destruction. On the contrary, we threaten them. In spite of the mindless and massive destruction of life being caused by human activity, the general pattern in living communities is one of a network of inter-dependencies.

More to the point, there are many species with which humans live in close relationships, and whose evolution we manage. These are the domesticated plants and animals that form the basis of our agriculture (cattle, rice), and who serve us as companions (dogs, cats, house plants). It is likely that our relationship with digital organisms will develop along the same two lines.

There will likely be carefully bred digital organisms developed by artificial selection and genetic engineering that perform intelligent data processing tasks. These would subsequently be "neutered" so that they can not replicate, and the eunuchs would be put to work in environments free from genetic operators. We are also likely to see freely evolving and/or partially bred digital ecosystems contained in the equivalent of digital aquariums (without dangerous peripherals) for our companionship and aesthetic enjoyment.

While this paper has focused on digital organisms, it is hoped that the discussions be taken in the more general context of the possibilities of any synthetic forms of life. The issues of living together become more critical for synthetic life forms implemented in hardware or wetware. Because these organisms would share the same physical space that we occupy, and possibly consume some of the same material resources, the potential for conflict is much higher than for digital organisms.

At the present, there are no self-replicating artificial organisms implemented in either hardware or wetware (with the exception of some simple organic molecules with evidently small and finite evolutionary potential [Nowick 1991][Feng 1992][Hong 1992]). However, there are active attempts to synthesize RNA molecules capable of replication [Beaudry 1992][Joyce 1992], and there is much discussion of the future possibility of self-replicating nano-technology and macro-robots. I would strongly urge that as any of these technologies approaches the point where self-replication is possible, the work be moved to specialized containment facilities. The means of containment will have to be handled on a case-by-case basis, as each new kind of replicating technology will have its own special properties.

There are many in the artificial life movement who envision a beautiful future in which artificial life replaces organic life, and expands out into the universe [Moravec 1988, 1989, 1993][Levy 1992a, 1992b]. The motives vary from a desire for immortality to a vision of converting virtually all matter in the universe to living matter. It is argued that this transition from organic to metallic based life is the inevitable and natural next step in evolution.

The naturalness of this step is argued by analogy with the supposed genetic takeovers in which nucleic acids became the genetic material taking over from clays [Cairn-Smith 1985], and cultural evolution took over from DNA based genetic evolution in modern humans. I would point out that whatever nucleic acids took over from, it marked the origin of life more than the passing of a torch. As for the supposed transition from genetic to cultural evolution, the truth is that genetic evolution remains intact, and has had cultural evolution layered over it rather than being replaced by it.

The supposed replacement of genetic by cultural evolution remains a vision of a brave new world, which has yet to materialize. Given the ever increasing destruction of nature, and human misery and violence being generated by human culture, I would hesitate to place my trust in the process as the creator of a bright future. I still trust in organic evolution, which created the beauty of the rainforest through billions of years of evolution. I prefer to see artificial evolution confined to the realm of cyberspace, where we can more easily coexist with it without danger, using it to enhance our lives without having to replace ourselves.

As for the expansion of life out into the universe, I am confident that this can be achieved by organic life aided by intelligent non-replicating machines. And as for immortality, our unwillingness to accept our own mortality has been a primary fuel for religions through the ages. I find it sad that Artificial Life should become an outlet for the same sentiment. I prefer to achieve immortality in the old fashioned organic evolutionary way, through my children. I hope to die in my patch of Costa Rican rain forest, surrounded by many thousands of wet and squishy species, and leave it all to my daughter. Let them set my body out in the jungle to be recycled into the ecosystem by the scavengers and decomposers. I will live on through the rain forest I preserved, the ongoing life in the ecosystem into which my material self is recycled, the memes spawned by my scientific works, and the genes in the daughter that my wife and I created.

2.17. Challenges

For well over a century, evolution has remained a largely theoretical science. Now new technologies have allowed us to inoculate natural evolution into artificial media, converting evolution into an experimental and applied science, and at the same time, opening Pandora's box. This creates a variety of challenges which have been raised or alluded to in the preceding essay, and which will be summarized here.

Respecting the Medium. If the objective is to instantiate rather than simulate life, then care must be taken in transferring ideas from natural to artificial life forms. Preconceptions derived from experience with natural life may be inappropriate in the context of the artificial medium. Getting it right is an art, which likely will take some skill and practice to develop.

However, respecting the medium is only one approach, which I happen to favor. I do not wish to imply that it is the only valid approach. It is too early to know which approach will generate the best results, and I hope that other approaches will be developed as well. I have attempted to articulate clearly this "natural" approach to synthetic life, so that those who choose to follow it may achieve greater consistency in design through a deeper understanding of the method.

Understanding Evolvability. Attempts are now underway to inoculate evolution into many artificial systems, with mixed results. Some genetic languages evolve readily, while others do not. We do not yet know why, and this is a fundamental and critically important issue. What are the elements of evolvability? Efforts are needed to directly address this issue. One approach that would likely be rewarding would be to systematically identify features of a class of languages (such as machine languages), and one by one, vary each feature, to determine how evolvability is affected by the state of each feature.

Creating Organized Sexuality. Organized sexuality is important to the evolutionary process. It is the basis of the species concept, and while remaining something of an enigma in evolutionary theory, clearly is an important facilitator of the evolutionary process. Yet this kind of sexuality still has not been implemented in a natural way in synthetic life systems. It is important to find ways of orchestrating organized sexuality in synthetic systems such as digital organisms, in a way in which it is not mandatory, and in which the organisms must carry out the process through their own actions.

Creating Multi-cellularity. In organic life, the transition from single to multi-celled forms unleashed a phenomenal explosion of diversity and complexity. It would seem then that the transition to multi-cellular forms could generate analogous diversity and complexity in synthetic systems. In the case of digital organisms, it would also lead to the evolution of parallel processes, which could provide us with new paradigms for the design of parallel software. The creation of multi-celled digital organisms remains an important challenge.

Controlling Evolution. Humans have been controlling the evolution of other species for tens of thousands of years. This has formed the basis of agriculture, through the domestication of plants and animals. The fields of genetic algorithms [Holland 1975][Goldberg 1989], and genetic programming [Koza 1992] are based on controlling the evolution of computer programs. However, we still have very little experience with controlling the evolution of self-replicating computer programs, which is more difficult. In addition, breeding complex parallel programs is likely to bring new challenges. Developing technologies for managing the evolution of complex software will be critical for harnessing the full potential of evolution for the creation of useful software.

Living Together. If we succeed in harnessing the power of evolution to create complex synthetic organisms capable of sophisticated information processing and behavior, we will be faced with the problems of how to live harmoniously with them. Given evolution's selfish nature and capability to improve performance, there exists the potential for a conflict arising through a struggle for dominance between organic and synthetic organisms. It will be a challenge to even agree on what the most desirable outcome should be, and harder still to accomplish it. In the end the outcome is likely to emerge from the bottom up through the interactions of the players, rather than being decided through rational deliberations.

2.18. Acknowledgements

This work was supported by grants CCR-9204339 and BIR-9300800 from the United States National Science Foundation, a grant from the Digital Equipment Corporation, and by the Santa Fe Institute, Thinking Machines Corp., IBM, and Hughes Aircraft. This work was conducted while at: School of Life & Health Sciences, University of Delaware, Newark, Delaware, 19716, USA, ray@udel.edu; and Santa Fe Institute, 1660 Old Pecos Trail, Suite A, Santa Fe, New Mexico, 87501, USA, ray@santafe.edu.

References

Adami, C., 1993, *Unpublished*, Learning and complexity in genetic auto-adaptive systems. Caltech preprint: MAP - 164, One of the Marmal Aid Preprint Series In Theoretical Nuclear Physics. Adami has used the input-output facilities of the new Tierra languages to feed data to creatures, and select for responses that result from simple computations, not contained in the seed genome (chris@almach.caltech.edu).

Anonymous, 1988, Worm invasion, *Science*, 11-11-88, 885.

Barton-Davis, P., *Unpublished*, Independent implementation of the Tierra system, contact: pauld@cs.washington.edu.

Beaudry, A.A., Joyce, G.F., 1992, Directed evolution of an RNA enzyme, *Science*, **257**, 635.

Bell, G., 1982, *The masterpiece of nature: the evolution and genetics of sexuality*, University of California Press (Berkeley).

Benner, S.A., Ellington, A.D., Tauer, A., 1989, Modern metabolism as a palimpsest of the RNA world, *Proc. Natl. Acad. Sci.*, **86**, 7054.

Brooks, R., *Unpublished*. Brooks has created his own Tierra-like system, which he calls Sierra. In his implementation, each machine instruction consists of an opcode and an operand. Successive instructions overlap, such that the operand of one instruction is interpreted as the opcode of the next instruction (brooks@ai.mit.edu).

Burstyn, H.L., 1990, RTM and the worm that ate internet, *Harvard Magazine*, **92**, 23.

Buss, L.W., 1987, *The evolution of individuality*, Princeton University Press (Princeton), 203.

Cairn-Smith, A.G., 1985, *Seven clues to the origin of life*, Cambridge University Press (Cambridge).

Carroll, L., 1865, *Through the Looking-Glass*, MacMillan (London).

Chao, L., Vargas, C., Spear, B.B., Cox, E.C., 1983. Transposable elements as mutator genes in evolution, *Nature*, **303**, 633.

Charlesworth, B., 1976, Recombination modification in a fluctuating environment, *Genetics*, **83**, 181.

Darwin, C., 1859, *On the origin of species by means of natural selection or the preservation of favored races in the struggle for life*, Murray (London).

Davidge, R., 1992, *Processors as organisms*, CSRP 250, School of Cognitive and Computing Sciences, University of Sussex. Presented at the ALife III conference (robertd@cogs.susx.ac.uk).

Davidge, R., 1993, Looping as a means to survival: playing Russian roulette in a harsh environment, *Self organization and life: from simple rules to global complexity*, proceedings of the second European conference on artificial life, (robertd@cogs.susx.ac.uk).

Davidson, E.H., Britten, R.J., 1979, Regulation of gene expression: Possible role of repetitive sequences, *Science*, **204**, 1052.

Dayhoff, J., 1990, *Neural network architectures*, Van Nostrand Reinhold (New York), 259.

DeAngelis, D., Gross, L, (eds), 1992, Individual based models and approaches in ecology, Chapman and Hall (New York).

de Groot, M., *Unpublished*, Primordial soup, a Tierra-like system that has the additional ability to spawn self-reproducing organisms from a sterile soup, (marc@kg6kf.ampr.org, marc@toad.com, marc@remarque.berkeley.edu).

Doolittle, W.F., Sapienza, C., 1980, Selfish genes, the phenotype paradigm and genome evolution, *Nature*, **284**, 601.

Eigen, M., 1993, Viral quasispecies, *Scientific American*, **269**, 32.

Farmer, J.D., Belin, A., Artificial life: the coming evolution, *Proceedings in celebration of Murray Gell-Mann's 60th Birthday*, Cambridge University Press.

Feferman, L., 1992, *Simple rules... complex behavior (video)*, Santa Fe Institute, Contact: fef@santafe.edu.

Feng, Q., Tae Kyo Park, Rebek, J., 1992, Crossover reactions between synthetic replicators yield active and inactive recombinants, *Science*, **256**, 1179.

Ghiselin, M., 1974, *The economy of nature and the evolution of Sex*, University of California Press (Berkeley).

Goldberg, D.E., 1989, *Genetic algorithms in search, optimization, and machine learning*, Addison-Wesley (Reading).

Gould, S.J., 1989, *Wonderful life*, W. W. Norton & Company, Inc., 347.

Gray, J., *Unpublished*, Natural selection of computer programs. This may have been the first Tierra-like system, but evolving real programs on a real rather than a virtual machine, and predating Tierra itself: "I have attempted to develop ways to get computer programs to function like biological systems subject to natural selection.... I don't think my systems are models in the usual sense. The programs have really competed for resources, reproduced, run, and "died". The resources consisted primarily of access to the CPU and partition space.... On a PDP11 I could have a population of programs running simultaneously", (Gray.James_L+@northport.va.gov).

Green, M.M., 1988, Mobile DNA elements and spontaneous gene mutation, M. E. Lambert, J. F. McDonald, I. B. Weinstein (eds.), Banbury Report 30, Cold Spring Harbor Laboratory, 41.

Halvorson, H.O., Monroy, A., 1985, *The origin and evolution of sex*, A.R. Liss (New York).

Hapgood, F., 1979, *Why males exist: an inquiry into the evolution of sex*, William Morrow (New York).

Hertz, J., Krogh, A., Palmer, R.G., 1991, *Introduction to the theory of neural computation*, Addison-Wesley Publishing Co. (Reading), 327.

Hogeweg, P., 1989, Mirror beyond mirror: puddles of life, *Artificial Life*, Santa Fe Institute Studies in the Sciences of Complexity, edited by C.G. Langton, Addison-Wesley (Redwood City), **6**, 297.

Holland, J.H., 1975, *Adaptation in natural and artificial systems: an introductory analysis with applications to biology, control, and artificial intelligence*, University of Michigan Press (Ann Arbor).

Hong, J.I., Feng, Q., Rotello, V., Rebek, J., 1992, Competition, cooperation, and mutation: improving a synthetic replicator by light irradiation, *Science*, **255**, 848.

Huston, M., 1979, A general hypothesis of species diversity, *American Naturalist*, **113**, 81.

Huston, M., 1992, Biological diversity and human resources, *Impact of science on society*, **166**, 121.

Huston, M., 1993, *Biological diversity: the coexistence of species on changing landscapes*, Cambridge University Press (Cambridge).

Huston, M., DeAngelis, D., Post, W., 1988, New computer models unify ecological theory, *Bioscience*, **38**, 682.

Jelinek, W.R., Schmid, C.W., 1982, Repetitive sequences in eukaryotic DNA and their expression, *Annual Reviews of Biochemistry*, **51**, 813.

Joyce, G.F., 1992, Directed molecular evolution, *Scientific American*, December 1992, 90.

Kampis, G., 1993, Coevolution in the computer: the necessity and use of distributed code systems, *ECAL93 proceedings*, Brussels, (gk@cfnext.physchem.chemie.uni-tuebingen.de).

Kampis, G., 1993, Life-like computing beyond the machine metaphor, *Computing with biological metaphors*, edited by R. Paton, Chapman and Hall (London), (gk@cfnext.physchem.chemie.uni-tuebingen.de).

Kauffman, S.A., 1993, *The origins of order, self-organization and selection in evolution*, Oxford University Press, 709.

Kerem, B., Rommens, J.M., Buchanan, J.A., Markiewicz, D, Cox, T.K., Chakravarti, A., Buchwald, M., Tsui, L., 1989, Identification of the cystic fibrosis gene: genetic analysis, *Science*, **245**, 1073.

Koza, J.R., 1992, *Genetic programming, on the programming of computers by means of natural selection*, MIT Press (Cambridge).

Langton, C.G., 1986, Studying artificial life with cellular automata, *Physica D*, **22**, 120.

Levy, S., 1992a, *Artificial Life, the quest for a new creation*, Pantheon Books (New York), 390.

Levy, S., 1992b, A-Life Nightmare, *Whole Earth Review #76*, Fall 1992, 22.

Litherland, J., 1993, *Open-ended evolution in a computerised ecosystem*, A Masters of Science dissertation in the Department of Computer Science, Brunel University, (david.martland@brunel.ac.uk).

Mac Arthur, R.H., Wilson, E.O., 1967, *The theory of island biogeography*, Princeton University Press, 203.

Maley, C.C., 1993, *A model of early evolution in two dimensions*, Masters of Science thesis, Zoology, New College, Oxford University, (cmaley@oxford.ac.uk).

Manousek, W., 1992, *Spontane Komplexitaetsentstehung – Tierra, ein Simulator fuer biologische Evolotion*, Diplomarbeit, Universitaet Bonn, Germany, (Kurt Stueber, stueber@vax.mpiz-koeln.mpg.d400.de).

Margulis, L., Sagan, D., 1986, *Origin of sex*, Yale University Press (Haven).

Marx, J.L., 1989, The cystic fibrosis gene is found, *Science*, **245**, 923.

Maynard Smith, J., 1971, What use is sex?, *J. Theoret. Biol.*, **30**, 319.

Maynard Smith, J., 1992, Byte-sized evolution, *Nature*, **355**, 772.

Maynard Smith, J., Dowson, C.G., Spratt, B.G., 1991, Localized sex in bacteria, *Nature*, **349**, 29.

Mead, C., 1993, *Analog VLSI and neural systems*, Addison-Wesley (Reading), 371.

Michod, R.E., Levin, B.R., 1988, *The evolution of sex: an examination of current ideas*, Sinauer Associates (Sunderland).

Moravec, H., 1988, *Mind Children: the future of robot and human intelligence*, Harvard University Press (Cambridge).

Moravec, H., 1989, Human culture: a genetic takeover underway, *Artificial Life*, edited by C.G. Langton, Santa Fe Institute Studies in the Sciences of Complexity, Addison-Wesley (Redwood City), **6**, 167.

Moravec, H., 1993, Pigs in cyberspace, *Extropy #10*, Winter/Spring 1993.

Morris, S.C., 1989, Burgess shale faunas and the Cambrian explosion, *Science*, **246**, 339.

Nowick, J., Feng, Q., Tijivikua, T., Ballester, P., Rebek, J., 1991, *Journal of the American Chemical Society*, **113**, 8831.

Orgel, L. E., Crick., F.H.C., 1980, Selfish DNA: the ultimate parasite, *Nature*, **284**, 604.

Rasmussen, S., Carsten, K., Feldberg, R., Hindsholm, M., 1990, The coreworld: emergence and evolution of cooperative structures in a computational chemistry, *Physica D*, **42**, 111.

Rasmussen, S., Knudsen, C., Feldberg, R., 1991, Dynamics of programmable matter, *Artificial Life II*, edited by C.G. Langton, C. Taylor, J. D. Farmer, & S. Rasmussen , Santa Fe Institute Studies in the Sciences of Complexity, Addison-Wesley (Redwood City), **10**, 211.

Ray, T.S., 1979, *Slow-motion world of plant "behavior" visible in rainforest*, Smithsonian, **9**, 12.

Ray, T. S., 1991a, An approach to the synthesis of life, *Artificial Life II*, edited by C.G. Langton, C. Taylor, J. D. Farmer, & S. Rasmussen , Santa Fe Institute Studies in the Sciences of Complexity, Addison-Wesley (Redwood City), **10**, 371.

Ray, T. S., 1991b, Population dynamics of digital organisms, *Artificial Life II Video Proceedings*, edited by C.G. Langton, Addison-Wesley (Redwood City).

Ray, T. S., 1991c, Is it alive, or is it GA?, *Proceedings of the 1991 International Conference on Genetic Algorithms*, edited by R.K. Belew & L.B. Booker, Morgan Kaufmann (San Mateo), 527.

Ray, T. S., 1991d, Evolution and optimization of digital organisms, *Scientific Excellence in Supercomputing: The IBM 1990 Contest Prize Papers*, edited by K.R. Billingsley, E. Derohanes, H. Brown, Athens, GA, 30602, The Baldwin Press, The University of Georgia.

Ray, T. S., 1992a, Foraging behaviour in tropical herbaceous climbers (Araceae), *Journal of Ecology*, **80**, 189.

Ray, T. S., 1992b, Tierra.doc, Documentation for the Tierra Simulator V4.0, 9-9-92, Newark, DE: Virtual Life, The full source code and documentation for the Tierra program is available by anonymous ftp at: tierra.slhs.udel.edu [128.175.41.34] and life.slhs.udel.edu [128.175.41.33], or by contacting the author.

Ray, T. S., 1994a, Evolution and complexity, *Complexity: Metaphors, Models, and Reality*, edited by G.A. Cowan, D. Pines and D. Metzger, Addison-Wesley Publishing Co., 161.

Ray, T. S., 1994b, Evolution, complexity, entropy, and artificial reality, *Physica D*, **75**, 239.

Rommens, J.M., Iannuzzi, M.C., Kerem, B. Drumm, M.L., Melmer, G., Dean, M., Rozmahel, R., Cole, J.L., Kennedy, D., Hidaka N., Zsiga, M., Buchwald, M., Riordan, J.R., Tsui, L.C., Collins, F.S., 1989, Identification of the cystic fibrosis gene: chromosome walking and jumping, *Science*, 245, 1059.

Riordan, J.R., Rommens, J.M., Kerem, B., Alon, N., Rozmahel, R., Grzelczak, Z., Zielenski, J., Lok, S., Plavsic, N., Chou, J.L., Drumm, M.L., Lannuzzi, M.C., Collins, F.S., Tsui, L.C., 1989, Identification of the cystic fibrosis gene: cloning and characterization of complementary DNA, *Science*, **245**, 1066.

Skipper, J., 1992, The computer zoo: evolution in a box, *Toward a practice of autonomous systems*, Proceedings of the first European conference on Artificial Life, edited by F.J. Varela and P. Bourgine,. MIT Press (Cambridge), 355, (Jakob.Skipper@copenhagen.ncr.com).

Sober, E., 1984, *The nature of selection*, MIT Press (Cambridge).

Spafford, E.H., 1989a, The internet worm program: an analysis, *Computer Communication Review*, **19**, 17, Also issued as Purdue CS technical report TR-CSD-823, (spaf@purdue.edu).

Spafford, E.H., 1989b, The internet worm: crisis and aftermath, *CACM*, **32**, 678.

Stanley, S.M., 1973, An ecological theory for the sudden origin of multicellular life in the late precambrian, *Proc. Nat. Acad. Sci.*, **70**, 1486.

Stearns, S.C., 1987, *The evolution of sex and its consequences*, BirkhŠuser Verlag (Boston).

Strickberger, M.W., 1985, *Genetics*, Macmillan Publishing Co. (New York).

Strong, D.R., Ray, T.S., 1975, Host tree location behavior of a tropical vine (Monstera gigantea) by skototropism, *Science*, **190**, 804.

Surkan, A., *Unpublished*, Self-balancing of dynamic population sectors that consume energy, Department of computer science, UNL, "Tierra-like systems are being explored for their potential applications in solving the problem of predicting the dynamics of consumption of a single energy carrying natural resource", (surkan@cse.unl.edu).

Syvanen, M., 1984, The evolutionary implications of mobile genetic elements, *Ann. Rev. Genet.*, **18**, 271.

Tackett, W., Gaudiot,J.L., 1993, Adaptation of self-replicating digital organisms, *Proceedings of the International Joint Conference on Neural Networks*, IEEE Press, (tackett@ipld01.hac.com, tackett@priam.usc.edu).

Taylor, C.E., Jefferson, D.R., Turner, S.R., Goldman, S.R., 1989, RAM: artificial life for the exploration of complex biological systems, *Artificial Life*, edited by C.G. Langton, Santa Fe Institute Studies in the Sciences of Complexity, Addison-Wesley (Redwood City), **6**, 275.

Thomas, C.A., 1971, The genetic organization of chromosomes, *Ann. Rev. Genet.*, **5**, 237.

Todd, P.M., 1993, Artificial death, *Proceedings of the Second European Conference on Artificial Life (ECAL93)*, **2**, 1048, Universite Libre de Bruxelles, (ptodd@spo.rowland.org).

Van Valen, L, 1973, A new evolutionary law, *Evolutionary Theory*, **1**, 1.

Williams, G.C., 1975, *Sex and evolution*, Princeton University Press (Princeton).

Chapter 3

Animated Artificial Life

Jeffrey Ventrella
jeffrey@ventrella.com
http://www.ventrella.com

3.1. Introduction

Life is motion. We see it stirring under blankets, shifting in a crowd, and sauntering down the street. We see it darting beneath leaves in a pond, scurrying around branches of a tree, and lumbering across a field. We can't always see what a living thing looks like in all its visual detail, but we can usually recognize its motion as something distinctly alive, as opposed to something moved by a machine or by the wind. Sometimes it's not just a matter of photoreceptors sewing together pixels to construct an image, but something incorporating brain and time, sensing a unique spatiotemporal event: life. What makes a thing appear alive? I think it might have something to do with a combination of apparent physical laws and some apparent adaptive behavior within that physical system, which is goal-directed. Living things have the quality of being tightly coupled with their environments–adapted, situated, immersed. This quality is achieved through recursive processes such as evolution and learning. The acquisition of life-like behavior in software can be achieved through similar means, and visualized as animated artificial life.

The discussions in this chapter are founded on the following premise: in designing real-time animated artificial life worlds, it is better not to begin by solving computer-graphics rendering problems in the classical way, and then to subsequently fill texturemapped polyhedra with behavior, as an afterthought. It is better to compose an ontology of dynamics first–biological and physical laws; to solve problems of adaptive behavior within this framework of biological and physical laws; and then to render them visually, in a way that speaks to the truth of the underlying dynamic. The simulations I will discuss throughout this paper were built with this in mind.

3.1.1. Disney Meets Darwin

The art of animation as invented by the Disney animators has been called "The Illusion of Life". Computer Animated cartooning is still catching up to the expressivity, humor, and life-filled motion possible with classic cel-based character animation, although recent animators such as those at Pixar have demonstrated some success in adapting computer technology to the fine art of classic character animation.

As film animators refine this marriage of new and old technologies, a new form of animated character emerges, not from within the film industry but from within various research labs. Add to Disney's "Illusion of Life", the "Simulation of Life", and witness a new technology–one which is compatible with the goals of Virtual Reality–future cyberspaces in which characters are not just animated, they are autonomous, reactive agents as well. Imagine further that these characters achieve their behaviors, in all their complexity and subtlety, by adapting to their environments and each other, "designing themselves", not through animation scripting, but through reproduction, crossover, mutation, and selection.

The surprises and novelty arising from such evolutionary systems can indeed be an annoyance and a distraction when the design objective is well understood, and when the animator knows exactly how a motion must look. But when seen as an art form or as a prototyping tool in which creativity, discovery, serendipity, and chance are welcome elements in the design process, these systems can be useful. A whole world of strange and funny animated behavior lies dormant, waiting to be fished out of the primordial soup.

3.1.2. Designing Emergence

According to one account, L. Frank Baum, the author of "The Wizard of Oz", was wandering around in a trance while developing a story, for several weeks. His wife asked what was wrong. He said that his characters wouldn't do what he wanted them to do. Days later, he seemed better and was busy back to work. His wife asked him how he had solved his problem. He said, "I let them do what they want to do" [American Heritage 1964]. Many creative projects benefit from this approach: a work in progress often gets weighted down by top-down design, too much intentionality from its inventor, and cries out to be set free–to acquire autonomy, to be what it naturally wants to be. While this may have little to do with most problems in scientific investigation, it does touch upon problems in synthesis and simulation, which are tools used in artificial life.

A key objective in artificial life projects is to construct systems in which self-organization and adaptive behaviors can arise spontaneously, not by design, but as emergent phenomena, having meaning in the context of the simulated environment. A duality is observed in the creation of artificial life worlds: while the goal is something emergent like self-organization or adaptability, the system is ultimately a designed artifact. Creating artificial life worlds, in this view, is an activity of *designing emergence*.

People often describe things within one particular level of detail, such as: the atomic, molecular, biological, psychological, ecological, social, etc. These levels overlap in our deeper pictures of reality. In composing an artificial life simulation, one chooses an arbitrary level as a base. On a "higher" descriptive level, life-like phenomena emerge as a result of the dynamics, when the simulation is run.

For instance, many artificial life simulations model populations of organisms whose spatial positions are represented by locations in a cellular grid, and whose physical means of locomotion (jumping from cell to cell) are not clearly specified, but simply represented as instantaneous relocations in space. The emergent properties in question may not require a deeper level of physical simulation than this. Physical mechanics are not needed. However, the experimenter may be more interested, for instance, in the specific motions of particular bodies than in general issues of population dynamics. A simulation in which an articulated body plan is integral to locomotion (as well as reproduction) requires another level of abstraction, and a deeper physical model becomes necessary, as a different set of emergent descriptors are sought.

In presenting my approach to Artificial Life, I will be describing a variety of experiments. I consider this body of work to be an exploration and a celebration of autonomously generated motion and form. The original impetus is not biological science, although biological principles have been incorporated into the work. The origins are in the study of aesthetics, perception, and creative design, particularly in the context of time-based media.

Towards the quest for increasing autonomy, and more interesting emergent phenomena, I have been progressively deepening my artificial life models. These include a variety of simply-rendered articulated figures incorporating physics and periodic motor control. The most recent simulation consists of a population of many hundreds of swimming creatures which animate in real-time. The creatures come in a large variety of anatomies and motion styles. They are able to mate with each other, and choose who has the "sexiest" motions, enabling evolution of swimming locomotion and anatomy which is attractive (beauty of course being in the eye of the beholder). The best swimmers reproduce their genes (because they can swim to a mate), and the most "attractive" swimmers get chosen as mates. In this simulation, not only are the aesthetics of motion and form subject to the chaotic nature of genetic evolution, but the creatures themselves partake in the choice of what forms and motions emerge. Emergent phenomena are observed on a number of levels, from population dynamics down to details as small as the idiosyncratic wiggling of a body part for sexual attraction.

3.1.3. Organization of This Chapter

This chapter describes an approach to the new science of Artificial Life [Langton, 1989], stemming from animated art and a design process for creating it. It is an expansion of a paper entitled "Designing Emergence in Artificial Life Worlds" [Ventrella 1998], which was presented at the *Virtual Worlds* Conference in Paris in 1998.

The simulations are presented in chronological order, although a bit of shifting around has been done in order to introduce topics in a more logical way. In these simulations, emergent behavior is enabled by progressively designing levels of autonomy into the model. In a very special way, the methodology itself has evolved over time. The simulations are used to illustrate how this methodology was built, with the first examples being quite primitive, and with the most recent simulation incorporating deeper physics and more autonomy.

Section 1 introduces the main concepts of this paper. Section 2 discusses background research and similar approaches to animated artificial life. Section 3 comprises the bulk of this paper, and provides a tour of a series of artificial-life-based experiments, accompanied by commentary. Section 4 introduces the latest set of simulations, in which organisms mate autonomously. Section 5 discusses the issue of rendering. Section 6 discusses various topics pertaining to artificial life and its relation to artificial intelligence. Finally, a conclusion is given in section 7.

3.2. Background

The spirit of animated artificial life existed before computers, in the form of wind-up toys, mechanical automatons, zoetropes, and Rube-Goldbergian carnival attractions. These physical artifacts are, by necessity, designed in a mostly top-down manner. Robotics research has begun to incorporate learning and evolutionary computation in the control of physical machinery, so that behaviors can adapt to complex unpredictable environments–bottom up design. Architectures that exhibit *emergent functionality* [Maes 1990], through intensive interaction with a dynamic environment, demonstrate this design approach. It can be argued that if any form of artificial life is to be considered "alive", its environment must be real, not virtual. It must be physical and truly situated in the world, no smoke and mirrors.

Though less able to pass the "is it alive" test, computer simulations occurring in virtual spaces of computer memory enable greater artistic flexibility, and supply a context for basic research in adaptive behavior. A number of physically-based locomotion systems have been designed for animation utilizing forward dynamics, rather than kinematic methods, for achieving realistic animal motion [McKenna 1990] [Raibert 1991]. Early examples of using artificial life principles in computer animation include *Boids* [Reynolds 1987], in which collective behaviors (flocking, herding, etc.) emerge from many interacting agents. [Badler 1991] describes physically-based modeling techniques and virtual motor control systems inspired by real animals used to automate many of the subtle, hard-to-design nuances of animal motion. In task-level animation [Zeltzer 1991] and the space-time constraints paradigm [Witkin 1988] these techniques allow an animator to direct a character on a higher level.

3.2.1. Genetic Algorithms in Animation

Models of *Creation* are made implicit every time an artist or engineer designs a body of work and establishes a methodology for its creation. Darwinian models are different (though not absent of a creator). In Darwinian models, important innovation is relegated to the artifact. Invention happens after initial design. Surprises often result. Genetic algorithms [Holland 1975][Goldberg 1989] are designed to take advantage of lucky mutations. They are serendipity engines. Genetic algorithms have also been shown to enhance the non-linear design subprocesses of "explore, evaluate, and refine", as described by [O'Reilly 1998], for architectural design. More relevant is the notion that genetic algorithms may be good candidates as assistants for prototyping in the design of animated motion [Ventrella 94a].

The genetic algorithm has been used for evolving goal-directed motion in physically-based animated figures, including a technique for evolving stimulus-response mechanisms for locomotion [Ngo 1993]. A holistic model of fish locomotion with perception, learning, and group behaviors, which generates beautifully realistic animations, was developed by Terzopoulos, Tu, and Grzeszczuk [Terzopoulis 1994].

The virtual creatures of Sims [1994 a,b] are the most celebrated examples from the growing zoology of artificial life entities entering the scene. Sims developed a comprehensive model for defining a creature genetically, using the genetic programming paradigm [Koza 1992]. Sims's virtual creatures incorporate deep physics and 3D shaded modeling, and are an excellent application for the Connection Machine, a massively parallel technology, for representing a massively parallel phenomenon–evolution.

3.3. A Collection of Artificial Life Experiments

3.3.1. Computer System

The creatures described in this paper were developed through a series of experiments with a somewhat different approach than many other experiments of the new Frankensteinian Order: importance is placed on real-time animation, and user interaction with the creatures, much like in a computer game. Because of this emphasis, *lighter* physics models are designed, and graphical representation is expressed in a more iconic, economical form, allowing for higher frame-rate and tighter interaction. To this end, an architecture was designed for fast physics and graphics rendering. A typical simulation/animation loop is illustrated in figure 1.

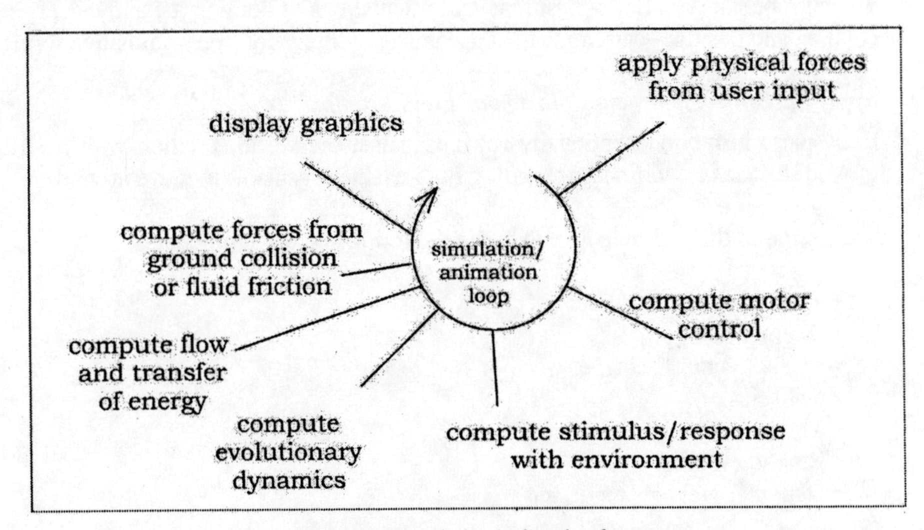

Figure 1. Simulation/animation loop.

The steps illustrated are not necessarily performed in the order shown. User interface control is not included here. All of the steps are performed exactly once per cycle. Simulation time is updated by units of 1 each cycle, and animation frame rate is equal to simulation speed. The steps are described below.

Apply Physical Forces from User Input

Mouse and keyboard events can be mapped to physical events in the simulated world to enable interaction with creatures as well as manipulation of a viewpoint (2D or 3D camera).

Compute Motor Control

Autonomous mechanisms cause body parts of creatures to move.

Compute Stimulus/Response

Body part motions and mental states are changed by environmental stimuli.

Compute Flow and Transfer of Energy

Energy is exchanged between the environment and organisms, and supplies the fuel for the creatures to live.

Compute Evolutionary Dynamics

Creatures can mate, reproduce with mutation, and die, either a standard genetic algorithm, or through "artificial natural selection" (no explicit fitness function).

Display Graphics

The current state of affairs–positions and orientations of body parts, and camera orientation and position–determine the appearance of things for every animation frame.

Apply Forces from Ground Collision or Fluid Friction

Body parts in motion continually rub up against the world. Friction with a virtual ground surface or a virtual surrounding fluid enable organisms to move around.

The design of these simulations includes the following key components:

- a morphological scheme
- an embryology
- a motor control scheme
- a physics
- a nervous system
- genetic evolution
- computer graphics rendering

These are each explained in the sections below.

3.3.2. Walkers

In "Oh Superman", performance artist Laurie Anderson describes walking as a continual process of catching oneself while almost falling, by placing one foot out in the direction of the fall, stopping the fall, and then almost-falling again, only to be stopped by the alternate foot [Anderson 1981]. This notion inspired the idea to construct a 3D bipedal figure which has mass and responds to gravity and which is always on the verge of falling over and so must continually move its legs in order to keep from falling, to stay upright, and to walk.

The first project in the series began an attempt to build a 3D stick figure which could stand up by way of internal forces bending various joints. The original idea was to model a human-like form, but this was too large a task at the time. To simplify the problem, the bipedal figure was reduced to the simplest anatomy possible which can represent bipedal locomotion: two springy jointed sticks, called "Walker", shown in figure 2. It could be argued that there can be no "walking" without "knees." But then again, the term, "walking" is used only in the qualitative sense here, as are the terms "head", and "feet", hence the preponderance of quotes around names for body parts and functions in this paragraph.

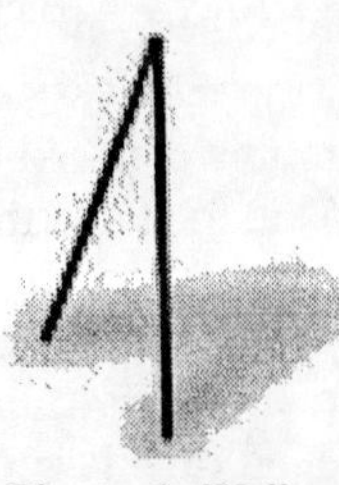

Figure 2. Walker.

A Walker's head experiences a pull in the direction of its goal (a virtual plate of food on the ground), proportional to the Walker's distance from the goal. This causes the Walker to tilt in the direction of the goal (an inaccurate model of the desire to be somewhere else, yet sufficient for this exploration). A Walker can perceive how much its body is tilting. This perception is used to modulate the motions in its legs (in a genetically determined way). As its head is pulled towards its goal, its legs move in such a way as to potentially keep it upright and afford locomotion. Internal forces cause the legs to rotate about the *head joint*, in periodic fashion, using sine functions. Parameters in the sine functions for each leg, such as amplitudes, frequencies, and phases, are genetically determined, varying among a population of Walkers, and are allowed to evolve over time within the population. Responses to body tilt are also genetically determined.

To optimize these parameters for walking, the population of Walkers is continually updated, generation-by-generation, with the best Walkers giving birth to new Walkers, and the worst ones dying off. At the evolutionary start, the motions of most of the Walkers are awkward and useless, with much aimless kicking, and so most of the Walkers immediately fall as soon as they are animated. Due to genetic variation, some move their legs such that they can stay upright for slightly longer periods, as they are pulled towards the food. These Walkers have higher fitness values. By using a simple genetic algorithm which repeatedly tests for fitness and reproduces the most fit Walkers in the population, with small random genetic mutations, the population improves locomotion skill over time, while maintaining slight variation throughout evolution. The genetic algorithm scheme used for Walker and subsequent simulations is described in more detail in section 3.3.13.

3.3.3. Food that Runs Away From You Keeps You Fit

A special feature is included in the simulation to help optimize locomotion: the virtual plate of food which causes the attractive force for the Walker also supplies energy to sustain the Walker for a specified duration of time. A Walker's energy level is continually decreasing. If it reaches zero, the Walker stops kicking and it falls over. So, if a Walker reaches critical proximity to the food and thus "partakes", its energy level is increased by a specified amount. Furthermore, the food possesses a minimal degree of adaptive evasive behavior. If a Walker partakes of the food, the food begins to move away from the Walker, experiencing a repulsion force analogous to the attractive force pulling the Walker to the food, only proportional to the inverse square of the distance from the Walker. Each time the Walker partakes, the food's repulsion force increases, causing it to move away more quickly as the Walker approaches. The back-and-forth adaptation resulting from this creates increasing demand for faster locomotion in order for a Walker to get to the food and thus to maintain energy.

This primitive form of co-evolution provided for Walkers adapted to turning on a dime, while pursuing a constantly moving target. To demonstrate Walker's ability to react to a constantly changing situation, a virtual "leash" was attached to the head, which a user could gently pull around (this is shown in figure 3, illustrated with motion blur). Also, scenarios were constructed in which multiple Walkers had attractions and repulsions to each other, generating novel pursuit and evasion behaviors.

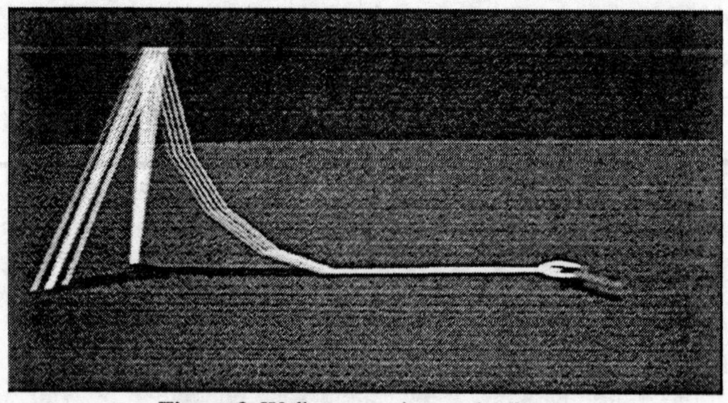

Figure 3. Walker on an interactive leash.

3.3.4. Springy Issues

Walker is the first in a series of artificial life creatures in this system consisting of linked body parts which are connected at their ends and which can rotate about their joints. Walker bodies are based on a spring model: there are three point masses: head, foot 1, and foot 2, each whose velocities are affected by gravity, ground friction, and spring forces. Spring forces are used to keep the feet and head close to an ideal distance (leg length). Simulation of rotations of legs about the head joint are achieved by incremental translational forces in the feet, oriented tangent to a sphere whose center is located at the head. The design of Walker had a number of flaws, one of which was the fact that, being a pair of joined springs instead of a rigid body model, there was no way to identify an explicit orientation of the body, and so rotational control was without context to a local body coordinate system. It required a physics hack. But despite these holes in the physics model, there was clearly something which adapted to meet its goals within that physical system, and it generated sufficiently life-like locomotion behavior. This early experiment provided a useful context for further exploration.

3.3.5. Other Walkers

The next stage in the simulation used fixed-length segments, and shifted geometry down to two dimensions. Figure 4 shows a creature from this later simulation of 2D creatures, consisting of five rigid interconnected segments, with four effective joints. As in the case of Walker, rotations about the joints are oscillatory, controlled by sine functions with parameters determined by genes. In this scheme, rotation is explicit and results in deterministic transformations of geometry. There is no food reward in this simulation. Fitness in this scheme is measured simply as distance traveled from the starting position during an animation sequence of specified duration. Even though creatures in this population are modeled in 2D, locomotion is possible, with evolved populations consisting of creatures ambulating either to the left or to the right. A distinctive feature of evolved locomotion in these populations is the emergence of phase offsets in joint angle oscillations converging close to 90 degrees, which correspond to the characteristic alternating gaits in many limbed animals.

Figure 4. 2D walking figure.

3.3.6. Morphology

It is one thing to link together a collection of segments of predetermined length in a predetermined topology and then allow the motions of the joints to evolve for the sake of locomotion. It is another thing entirely to allow morphology itself to evolve.

In the next simulation in this artificial life series, dimensionality was brought back up to three, and a morphological scheme was designed which allowed the number of segments, lengths, branching angles, and branching topology to vary. All of these factors are genetically determined. An important principle in this newer morphological scheme is that in this case, there are no implied heads, legs, arms, or torsos, as shown in figure 5. Creatures are open-ended structures to which Darwinian evolution may impart differentiation and afford implicit function to the various parts. In this scheme, any body part can become a "foot".

Figure 5. 3D creatures with variable morphology.

3.3.7. Embryology

Each creature possesses a series of genes organized in a template (the genotype). Genotypes are fixed-length arrays of real numbers. Each gene is mapped to a real or integer value which controls some specific aspect of the construction of the body, the motion of body parts, or reactivity to environmental stimuli. Genotypes get expressed into phenotypes through this embryological scheme. For most of the simulations discussed, the typical genotype for a creature includes genes for controlling the following set of phenotypic features:

- number of body parts
- frequency of sine wave rotations for all parts
- global control parameters affecting the following local features:
    - the parent of each part (determining branching topology)
    - the angle at which each part branches off its parent part
    - length of each part
    - amplitude of sine wave rotation in each part
    - phase offset of sine wave rotation in each part
    - amplitude modulator of sine wave (reactivity to continuous stimuli)
    - phase modulator of sine wave (reactivity to continuous stimuli)

For the 3D simulations, the phenotypic features above which pertain to angles are each represented by up to three genes, corresponding to three axes of rotation.

3.3.8. Lessons from the Blind Watchmaker

The design of embryologies has been one of the most interesting components in building artificial life worlds in this scheme. It is not always trivial to reduce the notion of a prolific body plan or motor control scheme to an elegant, data-compressed representation. [Dawkins 1989] promotes the concept of a "constrained embryology", whereby global aspects of a form (such as bilateral symmetry, segmentation, or fractal-like self similarity) can be represented with a small set of descriptive parameters. A representation with a small set of genes which are able to express, through their interactions during embryology, a very large phenotypic space, is desirable, and makes for a more evolvable population.

I have experimented with a variety of embryologies inspired by Dawkins's' explanations. In one simulation, I used a scheme which in fact goes completely against Dawkins's constrained embryology philosophy. In *Gene Pool* (section 2.4.4), each gene in the genotype controls only one local aspect of the phenotype. It is a fine-grained embryology with little innate structure. Consequently, the number of genes had to be made large (to account for prolific morphology and motion). This is not ideal for swift evolutionary search. On the other hand, using this scheme afforded the opportunity to identify emergent structure in phenotypes which originated from adaptation, with no influence from innate structure in embryological development. For instance, in *Gene Pool*, many emergent locomotion strategies appeared to utilize forms of symmetry. Since no structure for symmetry was designed into embryology, symmetry was observed as a desirable state that the population converged upon, purely through adaptation, without any genetic disposition.

The above example is an exception to the rule. In general, I have found that some implicit structure in embryology is useful. Furthermore, from a design research standpoint, building pregnant structure into tight representations provides a fascinating creative occupation.

3.3.9. Sims's Directed Graphs

The embryology used in Sims's virtual creatures employs an open-ended, flexible scheme which imparts a healthy amount of structure on morphological development, yet doesn't overly constrain embryology, as evidenced by the spectacular zoo generated by his system. In this scheme, the genotype is represented as a directed graph allowing for topologies and geometries with recurrent features. During development, signals are passed around through the accumulating hierarchy of body parts, and adjusted in the process. The recurrent nature of this scheme allows body parts to recursively *express themselves* with modifications. This corresponds to the existence of self-similarity in animal form and motion.

3.3.10. Motor Control

The simulations discussed here incorporate a straightforward motor control scheme. Each creature possesses a virtual motor which drives all the body parts to rotate in pendulum fashion, about their joints, repeatedly and continuously, from birth to death. Each body part can have its own amplitude and phase, as shown in figure 6. This motor control scheme roughly models periodic forms of locomotion in animals, in which the environment is relatively free of irregularities or distractions. It does not address motor control mechanisms for navigating irregular spaces, requiring stimulus/response mechanisms for moment-by-moment reactivity.

These periodic motions are further modulated by environmental factors, as explained in section 3.4.3, concerning the most recent simulation.

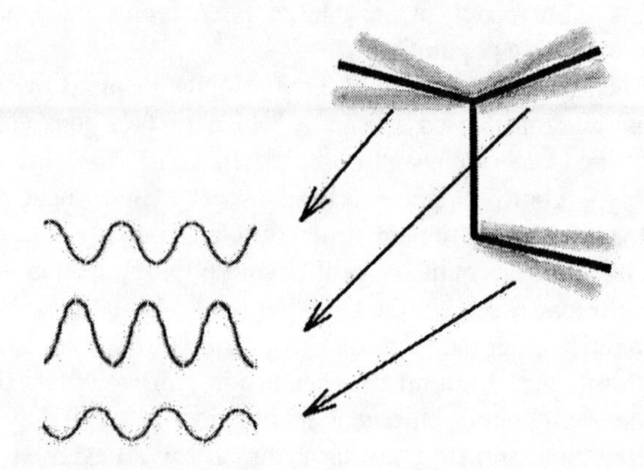

Figure 6. Different phases and amplitudes for bending body parts.

3.3.11. Physics

These simulations evolved from a simple animation technique whereby motion was made progressively more complex and autonomous. The notion of a physics also became incorporated in progressive steps. The early animation experiments that lead up to the current artificial life simulations were based on Koch curve fractal generators [Mandelbrot 1977]. These fractal curves are made of line segments connected at various angles, the sequence of which is recursive. Motion was achieved by simply changing angles in the data-compressed fractal generator incrementally over time, resulting in highly organic motion in the "data-amplified" fractalized forms. These motions are beautifully chaotic and organic. In viewing these abstract art pieces, parts of the fractals were found to move completely out of view as the animation progressed. As an experiment to keep these forms roughly centered on the computer screen, a simple feedback mechanism was incorporated, which computed the "center of mass of the object" and caused the form to gravitate towards the center of the screen in a loose, springy manner. Other techniques were added which, while being purely aesthetic visual treatments, pushed the animations closer to physical models, treating the fractal form as a complex jointed object in space rather than an image of recursion. These various techniques later became useful for modeling figures composed of line segments attached at angles, where the line segments were assumed to be skeletal components.

It may seem unconventional to progressively add physics to enhance animated motions. Often people speak of a computer model as being physically based or not being physically based. Is there an in-between? I believe that there is. In fact, there is no such thing as totally "real" physics. Every physical simulation is an approximation. The evolution of physics is a fundamental aspect of the methodology by which these artificial life simulations have been built. As the physical models deepen, so do the adaptations of the creatures.

3.3.12. Current Physics Scheme

In the present scheme, time is updated in discrete steps of unit 1 with each time step corresponding to one update of physical forces and one displayed animation frame. A typical creature is modeled as a collection of hierarchically-connected segments in a tree-like branching topology. It can change the angles of segment joints by small increments over time, resulting in pendulum-like rotations over extended time. The segment-bending procedure is borrowed from the fractal morphing technique described above.

Creatures are assumed to have infinite strength to bend joints without resistance from outside forces. A creature's body has position, orientation, linear velocity, and angular velocity. The body's position and orientation are updated at each time step by adding the body's linear and angular velocities. These velocities are in turn changed at each time step by various forces. These include:

Gravity

Always adding to the vertical component of the body's linear velocity

Ground Collision (for terrestrial simulations)

When the endpoint of a body segment collides with the ground (a flat horizontal surface), the angular and linear velocities of the body are affected (giving rise to effects such as bouncing, scraping, sitting at rest, etc.).

Water Friction (for aquatic simulations)

The motions of body parts (instantaneous linear velocities of the midpoints of individual body segments) create resistance within a fixed frame of viscous fluid. The relationship between the orientation of each body segment and the direction of its motion determines the magnitude and direction of each component force. The accumulations of these forces are summed to determine resultant changes in the body's linear and angular velocities.

User Interaction

Pressing certain keys on a computer keyboard or clicking the mouse cursor over certain areas in the scene of the simulation are translated into various forces acted upon the creatures. These can be used to "pick up" or drag creatures, or to *stimulate* them, for a variety of purposes.

The more recent creatures have body parts of varying thickness and length, and so there are implied uneven distributions of mass on either side of each joint. This requires more realistic modeling of orientations and center of mass changing over time as a result of joints bending. Actual torques between bending parts is not directly modeled. The effects of moments of inertia are modeled using a technique which is computationally economical for real-time animation speed. This recent upgrade in physics was required to avoid anomalies such as "the tail wagging the dog" (and I don't mean that figuratively!)

With computer processing speeds increasing the current rate, it will become reasonable to design deeper physical models, in which forces are based on first principles, while still maintaining real-time interactive speeds (always a high priority in this series of simulations). Each body part could then be modeled with its own independent position, orientation, velocity, and angular velocity. The interactions among parts would give rise to the coherent features which have been modeled globally so far. But, as a matter of fact, and however, I expect the next physics upgrade to bypass rigid body dynamics altogether and to model creatures on a springs-and-point mass level.

3.3.13. Evolution

The simulations discussed above use a variation of the standard genetic algorithm. A population of creatures are initialized with random genotypes, then each genotype is expressed as a phenotype and animated for a specified duration to establish a fitness ranking (in most cases, this is determined by how far it has traveled from the position at which it started, when the animation is complete). After this initial fitness test, a cycle begins: for each step in the cycle...

1. Two individuals are chosen from the population at random, with chances of being chosen proportional to fitness ranking.

2. Their genotypes are randomly crossed to generate one offspring genotype consisting of alternating gene sequences from both parents.

3. The offspring genotype is subject to chance mutation.

4. The lowest fitness ranking genotype in the population is removed, and the new offspring genotype is added to the population.
5. The offspring genotype undergoes embryology to create a phenotype.

6. The offspring phenotype is then animated to determine its fitness ranking within the population.

This process repeats for a specified number of cycles. This technique avoids premature convergence and guarantees that overall fitness never decreases in the population; the most fit creatures are never lost, and the least fit creatures are always culled [Ventrella 1994b].

3.3.14. Moving "Ahead"

In one experimental embryological scheme, I imposed an artificial symmetry and segmentation to the body, and designated a special part as the "head" (shown with a black dot in figure 7). Although I knew quite well where their heads were, these creatures had no notion of what it means to have a head, and so this was ignored by evolution. In fact, many evolved populations consisted of highly mobile creatures who used these head parts, along with other parts of their bodies, like feet. Some locomotion strategies appeared as if they could be painful. So extra components were added to the fitness function, rewarding creatures with bonus fitness points for such behaviors as: 1) not clobbering the head on the ground, 2) holding the head relatively high, and 3) moving the head relatively little. Figure 7 shows an example of a creature from a population in which head-height was added as a component to fitness, along with distance traveled.

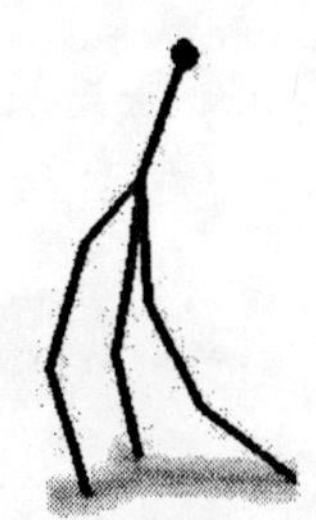

Figure 7. Evolutionary pressure for head-height.

The use of multiple fitness rewards pertaining to the head, compounded with the fundamental fitness reward for distance traveled, proved promising as a prototype for introducing a hint of top-down design in this evolutionary scheme. Encouraging lack of motion in the head, in addition to keeping it from hitting the ground, during evolution, resulted in some satisfying locomotion styles, such as the creature shown in four sequences of figure 8.

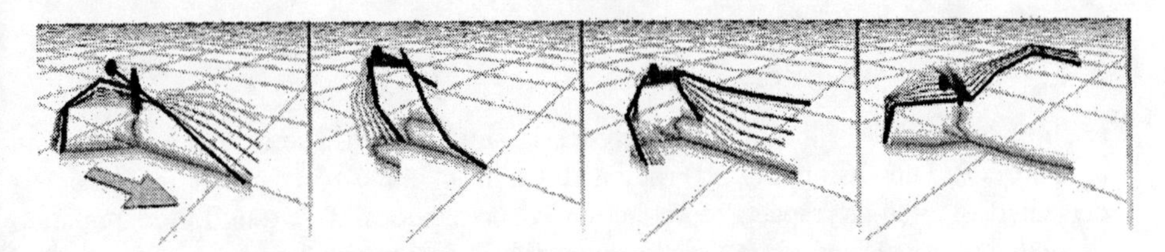

Figure 8. 3-legged locomotion strategy with steady head.

In this illustration, the creature is seen traveling in the direction indicated by the arrow in the first frame. Blurry afterimage lines are shown indicating the motion of the body. Notice how the head remains relatively steady while other body parts take care of most of the locomotion work.

3.3.15. Why Having No Inner Ear Means Having Three Legs

The creatures in these illustrations tend to use a tripod stance to stabilize themselves. Creatures in this scheme usually evolve 3-legged locomotion, sometimes 4. Bipedal locomotion does not evolve: this is probably due to the fact that they have no sensors for balance, and thus no way of adjusting to tilt (as Walker did). A future enhancement will be to add 3D "vestibular" sensors (analogous to the three semicircular canals in our inner ears), and adaptable reactivity to that sense, as a means (among others) to encourage evolution of bipedal locomotion.

3.3.16. Putting Meat on the Bones

An extension of this scheme includes bodies with variable thickness among parts. Figure 9 is a hand-drawn illustration of a representative 3D creature from this set. These creatures are the most complex of all the morphological schemes, and can exhibit locomotion styles and anatomies which exploit the effects of uneven distributions of mass among the body (consider for instance the way a giraffe uses the inertia of its neck when it runs).

Figure 9. A 3D creature with variable thickness in parts.

3.4. Sexual Swimmers

In the latest series of simulations, physics is shifted from a terrestrial model to an aquatic one. Dimensionality is brought down from three to two, but the model is deepened in another respect: reproduction is spontaneous. To ground the simulation more in a physical representation, the concept of a "generation" is removed. Instead of updating the entire population in discrete stages (a schedule determined by a "creator"), this model allows reproduction to occur asynchronously and locally among creatures able to reach proximity to each other in a spatial domain. The notion of an explicit fitness function for selection is removed [Ventrella 1996].

Whereas the previous simulations explicitly rewarded creatures for speedy locomotion, this simulation lets them reward themselves by mating. Speedy locomotion therefore becomes a prerequisite (along with many other behaviors), and not the sole objective.

3.4.1. The Energy Cycle

Energy exists in the fluid medium in ambient form, and is periodically converted into bits of food distributed randomly in the fluid. These food bits can be eaten by the swimming creatures to attain energy. They use this energy to move body parts and exert force against the fluid medium. Their energy levels decrease in proportion to physical exertion. Creatures are by default interested in mating, and pursue other creatures for this purpose. However, if a creature's energy dips below a specific threshold, it changes its mental state and pursues a nearby food bit instead of a mate. Swimming creatures who are genetically disposed to erratic motion waste large amounts of energy, and thus spend more of their lives pursuing food. They have less time to dedicate to propagating. Abnormally erratic creatures are usually not even able to reach a food source before losing all their energy, and so they usually die off early-on in evolution. Over evolutionary time, the most energy-efficient swimmers survive. It should come as no surprise that these swimmers, having evolved for energy-efficiency, are also beautiful to watch.

3.4.2. Physics and Genetics are Linked

It is important to point out that in this simulation, genetic evolution is totally reliant on the physical model which animates the creatures. Not only does it take Two to Tango, but they have to be able to swim to each other as well (more specifically, at least one has to be able to swim to the other). In this simulation, genotypes are housed in physically-based phenotypes, which, over evolutionary time, become more efficient at transporting their genotypes to other phenotypes, thereby enabling them to mate and reproduce their genes. This is the selfish gene at work. The emergence of locomotive skill becomes meaningful therefore in the context of the local environment: it becomes physically grounded, and the fitness landscape becomes dynamic.

Locomotion behavior evolves by way of local matings between creatures with higher (implicit) fitness than others in the local area. Often a maturing population will separate into two distinct regions in phenotype space, before it is taken over by one phenotype, corresponding to different areas of the fluid medium, where genotypes cannot readily mix.

Figures 10a and b illustrate two swimming styles which emerged from a simulation.

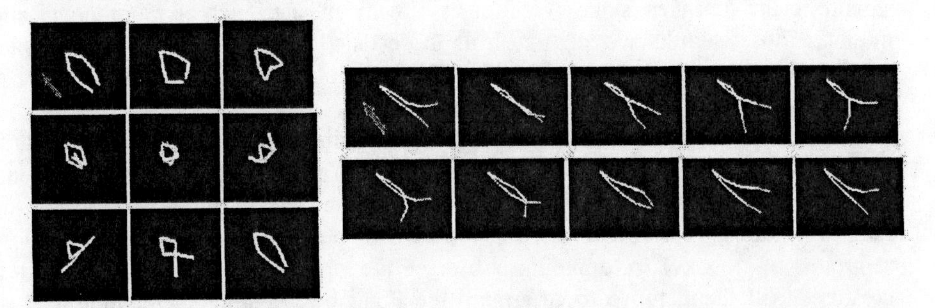

Figure 10. Sequences of images of swimming strategies: (a) paddling-style, and (b) undulating-style, which emerged through evolution.

3.4.3. Turn that Body

A swimming creature spends most of its life pursuing a potential mate. That potential mate may possibly be pursuing its own potential mate, and so it is often a moving target. A swimming creatures has to be able to continually orient itself in the direction it wants to go. A reactive system was designed which allowed a creature to detect how much it would need to turn its body while swimming. Essentially, the creature senses the angle between the direction of its goal and its own orientation as the stimulus to which it reacts, as shown in figure 11. This angular relation to the goal is dynamic, constantly changing as pursuer and pursued move throughout the fluid.

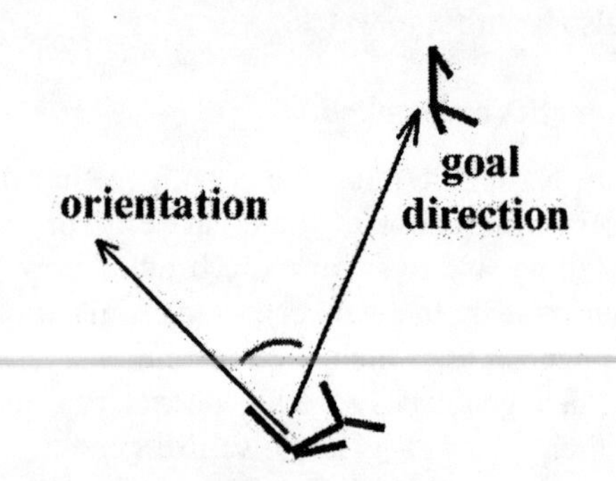

Figure 11. Turning stimulus.

As opposed to Walker, which experiences a "magic pull" in the direction of its goal, these creatures must use this sense to affect their own autonomous means of propulsion, using friction with the surrounding fluid.

Since the number of possible morphologies and motion strategies in these creatures is large, it would be inappropriate to top-down-design a turning mechanism. Since turning in the plane for a 2D articulated body can be accomplished by modulating the phases and amplitudes of certain part motions, it was decided that evolution should be the designer, since evolution is already the designer of morphology and motion. Thus, modulator genes (such as those listed in section 2.3.7) were implemented which can affect the amplitudes and phases in each part's motions in response to the stimulus.

A commercial product was derived from this simulation called "Darwin Pond" [RSG 1997]. It enables users to observe hundreds of swimming creatures, representing a large phenotypic space of anatomies and motions. They can be observed in the Pond with the aid of a virtual microscope, allowing panning across, zooming up close to groups of creatures, or zooming out to see the whole affair. One can create new creatures, kill them, move them around the Pond by dragging them, inquire their mental states, feed them, remove food, and tweak individual genes while seeing the animated results of genetic engineering in real-time. Figure 12 shows a screen shot of the basic Darwin Pond interface.

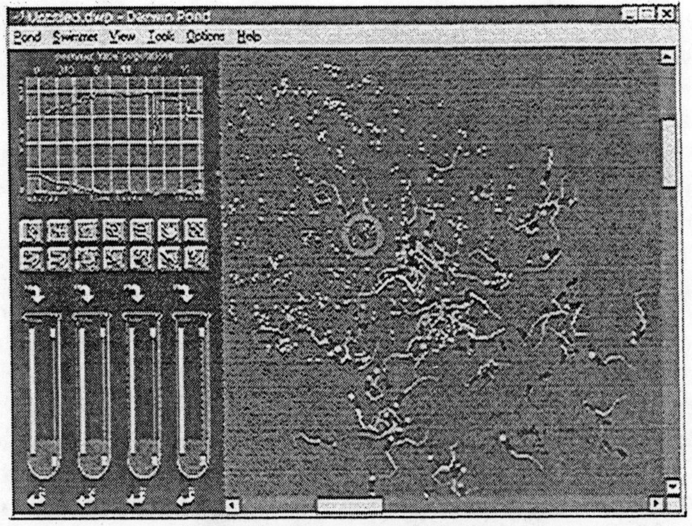

Figure 12. *Darwin Pond* interface.

A variation on this simulation, called "Gene Pool" was developed, and included a larger genotype-phenotype scheme, variable thickness in body parts, a deeper physics, and a conserved energy model. Figure 13 illustrates a collection of these 2D swimming creatures (before evolution has had any effect of body plan). They are called "swimbots".

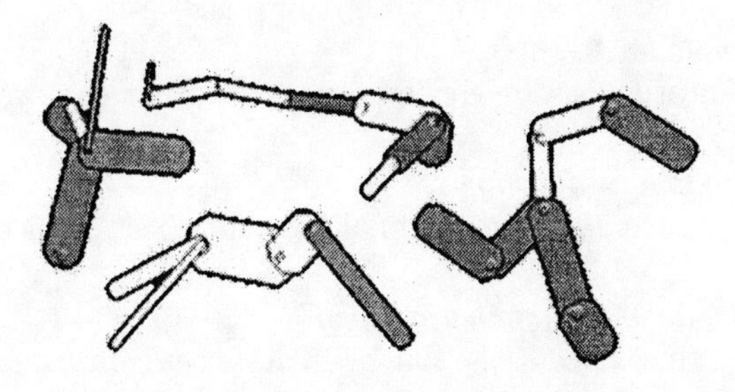

Figure 13. Swimbot anatomies.

3.4.4. What's Sex Got to Do With It?

Gene Pool deepens the simulation further by introducing mate choice in the process of reproduction. In this simulation, not only is the evolution of form and motion more physically grounded, and subject to the local conditions of the simulated world, but the aesthetic criteria which determine the "sexiest" motions also become a factor for reproduction.

What does "sexy" mean? Sexual selection in evolution is responsible for phenomena such as the magnificent peacock tail, and the elaborate colorful dances of many fish species, who sport curious body parts, adapted for mate attraction. Attraction-based features in many species may be so exaggerated and seemingly awkward that one would assume they decrease the overall fitness of the species, in which locomotion efficiency, foraging, and security are important. But the mere existence of these remarkable sexual features is proof that they are not fatal to the survival of the species. An interesting discussion on the topic of sexual selection and the biology of beauty can be found in [Moller 1997].

Computer models developed by Todd and Miller [1993], in which sexual selection drives a population of simulated organisms into arbitrary regions of phenotype space, supplied motivation to explore this intriguing side of evolution. A question was posed: could mate preferences for arbitrary phenotypic features inhibit the evolution of energy-efficiency in a population of locomotive creatures? For instance, if all creatures in the population were attracted to slow-moving bodies, and only mated with the most stationary creatures in the population, would locomotion skill still emerge? If so, what kind? What if they were all attracted to short, compact bodies? Would the body plan become reduced to a limbless form, incapable of articulated locomotion? If they were attracted to wild, energetic motion, would the resulting swimbots be burning off so much energy that most die before reproducing?

As an experiment, a simulation was built, including a variety of mate preference criteria determining which swimbots would be chosen [Ventrella 1998]. At the point of choosing a potential mate, a swimbot takes "snapshots" of each swimbot within its local view horizon, and then compares these snapshots and ranks them by attractiveness. It chooses the most attractive swimbot as its goal and then begins to pursue it. Criteria settings include:

- *attraction to long bodies*
  (length is measured as the greatest distance between any two body parts)

- *attraction to lots of motion*
  (motion is measured as the sum of all instantaneous speeds of all body parts)

- *attraction to bodies which are "open"*
  (openness is measured as the sum of all distances between parts)

- *attraction to massive bodies*
  (massiveness is measured as the sum of all body part areas)

The inverses of each of these phenotypic characteristics are also tested, making a total of eight attractiveness criteria. As expected, in simulations in which length was considered attractive, populations with elongated bodies with few branching parts emerged. In these populations, locomotion was accomplished by a tiny paddle-like fin in the back of the body, or by gently undulating movements. Figure 14 illustrates a local cluster of swimbots resulting from this simulation.

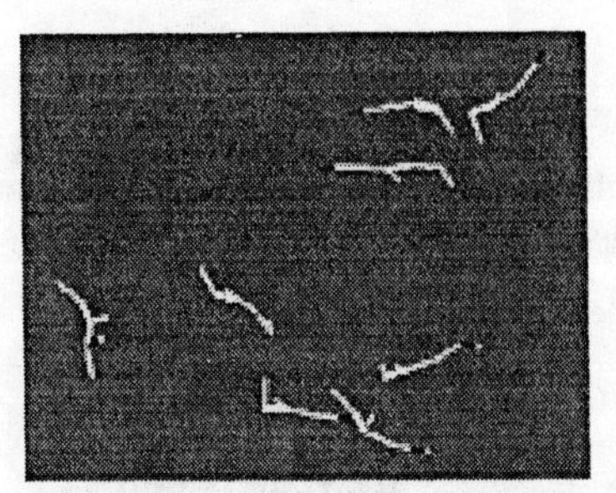

Figure 14. "Long is Beautiful".

Attraction to *open* bodies resulted in creatures which spent a large part of their periodic swimming cycles in an "open" attitude, as shown in figure 15. In this illustration, 13 images are shown in *Muybridge* form to illustrate approximately one swim cycle, sequenced from left-to-right. The creature is swimming in the direction towards its lower right. Notice that the stroke recovery (the beginning and end of the sequence) demonstrates an exaggerated "opening up" of the body. Having observed hundreds of swimming cycles in evolved populations, I can easily say that this is uncommon. It is likely that this behavior emerged as a by-product of the attractiveness criteria for the following reason: at the moment a creature is ready to take snapshots to size up potential mates within its view, the potential mates are all in various arbitrary stages of their swim cycles. So it is advantageous to be attractive as much as possible throughout the entire swim cycle. The more *open* a swimmer's body is, and the longer amount of time it is open throughout the swim cycle, the more *attractive* on average it will be.

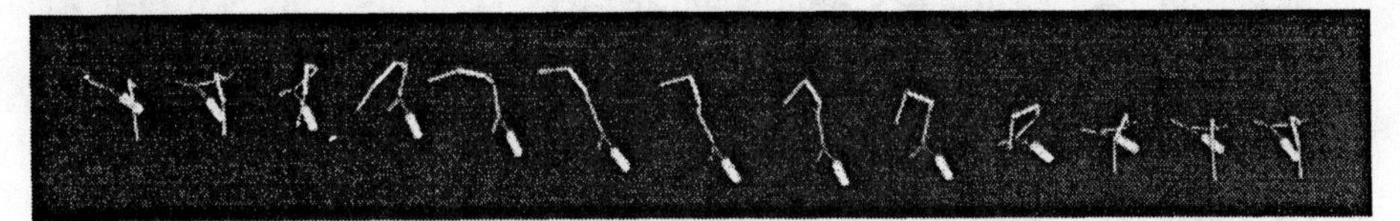

Figure 15. Swimbots which evolved through mate preferences for *open* bodies.

These experiments show how mate preference can affect the evolution of a body plan as well as locomotion style. The effects on energy-efficiency seem to be non-trivial: locomotion strategies emerge which sometimes take advantage of attraction criteria yet are still efficient in terms of locomotion. As usual with most artificial life simulations, the phenotype space has unexpected regions which are readily exploited as the population adapts–regions which a top-down designer might not think to visit.

3.4.5. Just Sit and Look Pretty

To be "fit" in this world doesn't necessarily mean to be fast, good at turning, or energy-efficient. It also includes being attractive. In fact, using some attractiveness criteria (such as lack of movement, or short bodies), it was found that populations could easily evolve to consist mostly of immobile creatures with hardly any anatomy. In such evolved populations, there are a few efficient swimmers who appear to be responsible for most of the mating. These individuals seem to be endowed with anatomy more associated with pursuit, while the majority of the individuals possess anatomies more attributable to attractiveness. This is a curious behavior. It appears that a form of polymorphism might be emerging in these situations, whereby a genotype is easily expressed in one of two distinct phenotypes.

3.4.6. Beauty Should Be in the Eye of The Beholder

For experimental purposes, the attractiveness criteria in these simulations were designed in a top-down, deterministic manner. To ground the simulation deeper, it would be better to allow beauty to be in the eye of the beholder and to be evolvable as well. This would require the design of pattern-recognition agents in creatures which size-up anatomies and motions of potential mates using their own genetically-determined criteria for beauty. These criteria should be sufficiently open-ended and variable to account for a very large space of possible criteria for beauty. Dawkins' concept of constrained embryology ideas come to mind again, only in this case, the method would be analytic instead of synthetic. Specific phenotypic features would be apprehended in a constrained way, through a "beauty filter".

3.5. Rendering

The invention called *image* remains a standard vessel for perceivable stuff which is measured as being either "realistic", "not realistic", or something in-between. This standard for which the adjective *realism* can be applied, has, in my opinion, obscured developers of new computergraphic media (like artificial life and computer games), in the pursuit of virtual realities. A different, and very potent, modality of realism can be rendered.

Some visually-oriented readers may never reach this section because throughout the paper there is a lack of rich visual illustration: the creatures are almost all stick figures. That is because the domain I have been exploring doesn't require any more than skeletal expressions of articulated form and motion. If these illustrations could be set to motion, however, one would immediately see where the depth lies: not simply in the x, y, or z dimensions, but in the four dimensions of space and time. One cannot see physics if there is no movement. Also, a realistic, deep physics can make up for a lack of visual detail. When this living motion is eventually clothed, the dynamism can still remain within, as determined by the physical and biological laws employed.

Part of the methodology here is to take advantage of something that the eye-brain system is good at: detecting movement generated by a living thing. Since it requires high frame rates for viewers to resolve physically-based motion, sacrifices need to be made to achieve this, due to available computer speeds. I have chosen not to spend much computational energy on heavyweight texturemaps, lighting models, and surface cosmetics, when more computation can be spent on deep physics, for animation speed (to show off the effects of subtle movements that have evolved). No pasting 2D images of Disneyesque eyes and ears on the creatures. Instead, as the simulation itself deepens (for instance, if light sensors or vibration sensors are able to evolve on arbitrary parts of bodies, as determined by evolution) primitive graphical elements visualizing these phenotypic features will be rendered. They may be recognizably eye and ear-like, or they may not. The important thing is to visualize what is there, rather than what is not. By not being obscured by cosmetics, the essential dynamics shows through.

3.5.1. Techniques Used

While I have chosen to compromise surface rendering for the sake of more direct visual expressions and faster animation, there are a few techniques worth mentioning, used to visualize the creatures.

For all the articulated stick figures, black 3D lines are drawn to represent body parts. Line occlusions are not dealt with: since the lines are all the same color, this is not a consideration. The interesting (and important) 3D visualization comes in the drawing a shadow. This is the most salient graphical element which gives the viewer a sense of the 3D geometry. The shadow is simply a "flattened-out" copy of the collection of body lines (without the vertical component and translated to the ground plane), projecting the figure vertically onto the ground. The shadow is drawn with a thicker linewidth, and in a shade slightly darker than the ground color. The shadow is drawn first, followed by the body part lines (painter's algorithm). In some simulations, a series of shadow objects are drawn in increasingly darker colors and increasingly thinner widths, to create a composite shadow object which appears to have blurry boundaries.

The progression in designing increasingly complex bodies is accompanied by a similar progression in rendering. When line-segment-based body parts were variably *fattened*, the thicknesses of the lines were likewise increased, such that they appeared like rectangles of varying widths, oriented along the lengths of the body parts. The swimbots in *Gene Pool* are 2D examples of this technique.

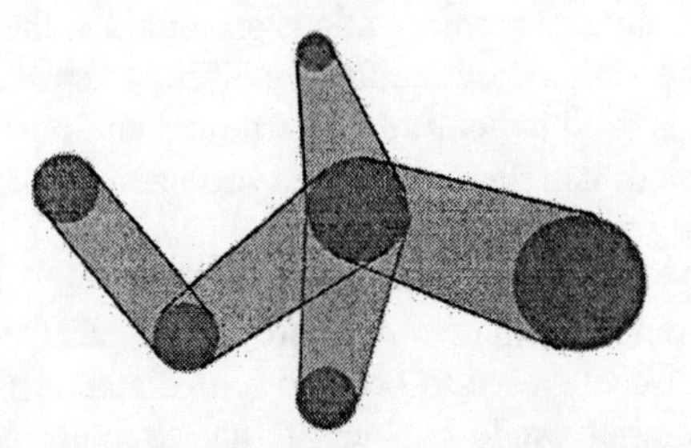

Figure 16. Representing body parts as spheres and cone sections.

In more advanced 3D bodies to be used in future simulations, each joint connecting body parts possesses a unique radius. The shape of a body part is determined by the radii of the connecting joints at either end of the part. Each body part is modeled as a portion of a cone connecting two spheres of different radii, as shown in figure 16. To render these parts, polygons oriented towards the viewpoint are drawn. Their shapes roughly correspond to the projections of the solid parts onto the viewplane.

Since spheres and connecting cones are able to be expressed in parametric form, it is not difficult to generate 2D polygons representing their projections. In this illustration, spheres are shown as disks of a darker color for clarification.

With this technique, one polygon per body part is drawn as a silhouette, along with an associated joint-sphere silhouette, instead of many polygons showing a faceted surface. This enables faster animation speeds. Custom shading techniques are currently being worked out at this time.

3.6. Discussion

3.6.1. Nervous Systems

It is possible that brains are local inflammations of early evolutionary nervous systems, resulting from the benefit in early animals to have *state* and generate internal models in order to negotiate a complex environment. This is an artificial life-friendly interpretation of brains.

The simulations I have described do not include extensive adaptive behavior in individual organisms. An enhancement in this design methodology would be to wire up the creatures with more connectivity and reactivity to the environment, and then to eventually plug-in evolvable neural nets (and the capability for them to grow) which complexify as a property of the creature's coupling with a complex environment. The Vehicles of [Braitenberg 1984] supply some inspiration for this bottom-up design approach.

This design philosophy does not consist of building a brain structure prior to establishing the nature of the organism's connectivity to the physical environment. If adaptive behavior and intelligence are emergent properties of biological life, then it is more sensible to design proto-bodies, and to wire up a proto-nervous system in which brain-like structures might emerge in direct response to the environment. The "Physical Grounding Hypothesis" [Brooks 1990], states that to build a system that is intelligent it is necessary to have its representations grounded in the physical world.

In the current system, there are hardly any brains to speak of. It includes a simple mental model which is (at this point) predetermined and purely reactive: sensors in a creature detect aspects of its relation to the environment, which can directly transition its mental state to another mental state. For instance, while in "looking for mate state", a creature will react to certain visual stimuli, which can cause it to enter into a new state, such as "pursuing mate". A future enhancement to this scheme includes neural material which can evolve to become a mediator of the sensor-actuator pathways. The neural material would evolve internal structure and representation in the context of the creature's direct coupling with the dynamic environment. So, for instance, instead of an environmental stimulus directly and unconditionally causing a predetermined change of mental state or body motion, the stimulus is processed along the way, resulting in many possible outcomes, depending upon some internal state and some environmental situations.

3.6.2. Muscles, Bones, and Nerves

Here is an idea for a possible artificial life project: imagine a simulation which models a simple environment, and uses three basic creature materials: muscles, bones, and nerves. Imagine concocting an embryology that uses a genotype to build structures (proto-creatures) out of these materials. Primordial, un-evolved proto-creatures would consist of chaotic assemblages in which longish bone elements are connected hinge-wise end-to-end. Muscles (bone-bending agents) would exist at bone connections. Nerves would consist of linear elements with neurons at their ends, wired up to bone ends as well as each other. These proto-creatures would be stimulated by internal motors and a few environmental triggers. Some of the nerve endings could be sensors that could respond to stimuli corresponding to the presence or positions of objects. This proto-creature idea is not unlike the method used by Sims [Sims 1994a] for the design of his virtual creatures' control systems.

Imagine further that many of the nerve endings are connective neurons–the connectionist stuff of neural nets. While some nerve endings could receive stimuli, and while other nerve endings could stimulate muscles to bend bones, many nerve endings would essentially become internal layers, and the embryology would be capable of constructing potentially many such neurons. This simulation would be capable of evolving neural-like structures which are three-dimensional and completely wedded to the "body" (muscles and bones), evolving physically along with the body, through selection. If one were to successfully build such a simulation, capable of evolution, would brain-like or central-pattern-generator-like nerve bundles naturally emerge in local areas of the body? Would nerve-bundles appear in crucial local areas of the body, where a little subsumtive computation might help? (cf. figure 17). If such brain-like structures were to emerge, this might serve as a nice illustration that the brain and body are not separated by a "Cartesian Lesion", but are co-evolved, and unified as a single *Bodymind.*

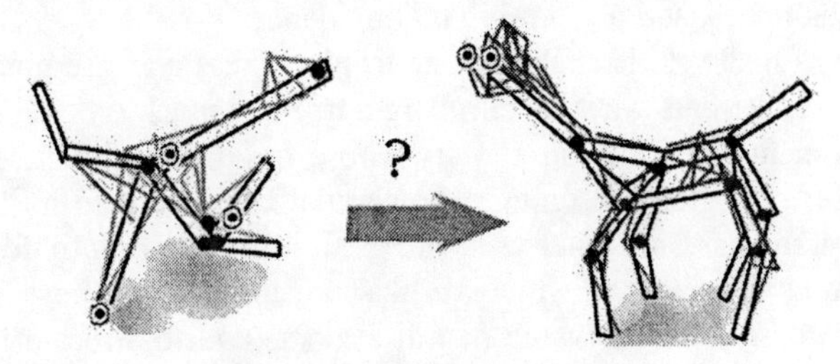

Figure 17. Emergent Brains.

3.6.3. Humans

We humans have always been extremely interested in *Homo sapiens*, and have rendered the likeness of this species in a myriad of ways throughout history. A recent motivation to render humans is accompanied by the arrival of a new communications medium: real-time 3D cyberspace for chat. Enter the *avatar*, a proxy for your identity; a computergraphic object representing you and your intentions. Since we are humans, we usually want our avatars to be humanoid, but of course they don't have to be. Humans are not only potent subject matter for art and communications media, they are the most familiar beings in our lives, and so our perceptions of things human are finely tuned. Humans are intensely responsive to faces and can read the most subtle of expressions; they are sensitive to skin tone, and can appreciate the well-crafted hues in a fine portrait; they are able to recognize familiar voices, and can easily detect mechanically-generated imitations; and finally, humans are superbly good at reading body language. Have you ever recognized a friend from a long distance away, simply from his or her distinctive gait? Body movement is one of the more unappreciated channels of expression and communication (especially now since person-to-person communication is so mediated by technology, and at such low bandwidth compared to the real thing).

While many of the human characters rendered in computer animations and computer games do impressively model key aspects of humans, they are almost always recognizable as synthetic. The successful Turing Test is a long way off. This might be because so much human modeling is simply a further extension of classic computer-graphic modeling techniques, whereby the surfaces of volumes are broken down into polygons, and shaded using a lighting model. Motion is achieved through techniques such as scripted keyframe animation or snippets of motion-capture. Motion-capture can achieve relatively good results, but how many pre-determined motion-captured scripts are required to represent any unique situation that a synthetic human is in at any given moment?

One reason that state-of-the-art virtual humans are unconvincing is that they are built "skin-first." If you slice one of these computergraphic models open, you will notice that it is hollow, whereas if you slice open a real human (and I am not suggesting you do this) you will notice a complex expression of billions of years of evolution in the biosphere, with a motor control system whose design history is punctuated by strategies for fleeing from predators, navigating rough terrains, leaping between tree branches, and performing mating dances.

It appears as if the skills enabling one to play chess and use grammar are a very recent evolutionary trend, whereas climbing a tree is a much older skill, perhaps more complex, and in fact possibly an activity whose basic cognitive mechanisms are the primordial stuff of higher reasoning. The simulations described in this paper do not aim to model humans. But if that were the goal, it may be best to design a vestibular system and an environment in which vertical, bipedal locomotion *might* be advantageous. Best to design an evolvable neural system (and to allocate lots of computer memory!) and a highly complex, dynamic environment, whereby something brain-like *might* billow as an internal echo of the external complexity. Best to design a simulation in which objects existing in the world *might* be exploited as tools and symbols, extended phenotypes, cultural memes. What might evolve then could be something more human-like than the wiggling forms shown in the illustrations of this paper. But in the meanwhile, there is no rush on my part to get to human levels with these simulations. And at any rate, imaginary animals can be very thought-provoking.

3.7. Conclusion

The artificial life worlds described here incorporate physical, genetic, and motor-control models which are spare and scaled down as much as possible to allow for computational speed and real-time animation, while still allowing enough emergent phenomena to make the experience interesting and informative. The effects of Darwinian evolution can be witnessed in less than a half-an hour (on a home computer). For educational and entertaining experiences, it is important that there be enough interaction, immersion, and discovery to keep a participant involved while the primordial soup brews. The average short-attention-span hardcore gamer may not appreciate the meditative pace at which an evolutionary artificial life system runs. But artificial life enthusiasts and artistically or scientifically oriented observers may find it just fine. For watching evolutionary phenomena, it sure beats waiting around a couple million years to watch lizards turn into snakes.

In this chapter, I have described many of the techniques, concepts, and concerns involved in a methodology for creating animated artificial life worlds. This approach does not begin with a biological research agenda but rather an eye towards the creation of forms and motions which are life-like and autonomous–an endeavor related to character animation. The artificial life agenda of studying emergent behavior by way of the crafting of simulations has become incorporated into this animation methodology, and with it, some key themes from evolutionary biology, such as sexual selection.

I hope that these ideas and techniques will be useful for further advances in animated artificial life, for education, entertainment, and research. I also hope that this and other such animated artificial life explorations will deepen our appreciation for the biosphere, and our kinship with all animals.

References

Anderson, L., 1981, *Oh, Superman* (phonograph record album).

Badler, N., Barsky, B., Zeltzer, D., 1991, *Making Them Move*, Morgan Kaufmann.

Braitenberg, V., 1984, *Vehicles: Experiments in Synthetic Psychology*, MIT Press (Cambridge).

Brooks, R., 1990, Elephants Don't Play Chess, *Designing Autonomous Agents*, edited by P. Maes, MIT Press (Cambridge), 3.

Dawkins, R., 1989, The Evolution of Evolvability, *Artificial Life*, edited by C.G. Langton, Addison-Wesley, 201.

Goldberg, D., 1989, *Genetic Algorithms in Search, Optimization, & Machine Learning*, Addison-Wesley (Reading).

Holland, J., 1975, *Adaptation in Natural and Artificial Systems*, University of Michigan Press (Ann Arbor).

Koza, J., 1992, *Genetic Programming: on the Programming of Computers by Means of Natural Selection*, MIT Press (Cambridge).

Langton, C.G. (ed.), *Artificial Life*, Addison-Wesley.

Maes, P., (ed.), 1990, Designing Autonomous Agents, MIT Press (Cambridge).

Mandelbrot, B., 1977, *The Fractal Geometry of Nature*, W. H. Freeman and Company, 34.

McKenna, M., Zeltzer, D., 1990, Dynamic Simulation of Autonomous Legged Locomotion, *Computer Graphics*, **24**, 29.

Miller, G., Todd, P., 1993, Evolutionary Wanderlust: Sexual selection with directional mate preferences, *From Animals to Animats II*, edited by Meyer, Roitblat & Wilson. MIT Press (Cambridge).

Ngo, T., Marks, J., 1993, Spacetime Constraints Revisited, *Computer Graphics*, 343.

O'Reilly, U., Ramachandran, G., 1998, A Preliminary Investigation of Evolution as a Form Design Strategy, *Artificial Life VI*, MIT Press (Cambridge).

Raibert, M., Hodgins, J.K., 1991, Animation of Dynamic Legged Locomotion, *Computer Graphics*, **25**, 349.

Reynolds, C., Flocks, Herds, Schools, 1987, A Distributed Behavioral Model, *Computer Graphics*, **21**.

RSG (Rocket Science Games, Inc., producer), 1997, *Darwin Pond* software product, (explained in the web site at: http://www.ventrella.com/Alife/Darwin/darwin.html).

Sims, K., 1994, Evolving Virtual Creatures, *Computer Graphics*, SIGGRAPH Proceedings, 15.

Sims, K., 1994, Evolving 3D Morphology and Behavior by Competition, *Artificial Life IV*, edited by R. Brooks & P. Maes, MIT Press (Cambridge).

Terzopoulos, D., Tu, X., Grzeszczuk, R., 1994, Artificial Fishes with Autonomous Locomotion, Perception, Behavior, and Learning in a Simulated Physical World, *Artificial Life IV*, edited by R. Brooks & P. Maes, MIT Press (Cambridge).

Ventrella, J., 1994a, *Disney Meets Darwin (An Evolution-based Interface for Exploration and Design of Expressive Animated Behavior)*, MIT Media Lab Master's Thesis, MIT Press.

Ventrella, J., 1994b, Explorations in the Emergence of Morphology and Locomotion Behavior in Animated Figures, Artificial Life IV, edited by R. Brooks & P. Maes, MIT Press (Cambridge).

Ventrella, J., 1996, Sexual Swimmers (Emergent Morphology and Locomotion without a Fitness Function), *From Animals to Animats*, MIT Press (Cambridge), 484.

Ventrella, J., 1998, Attractiveness vs. Efficiency (How Mate Preference Affects Locomotion in the Evolution of Artificial Swimming Organisms), *Artificial Life VI*, MIT Press (Cambridge).

Ventrella, J., 1998, Designing Emergence in Animated Artificial Life Worlds, *Virtual Worlds*, Lecture Notes in Artificial Intelligence, edited by J.C. Heudin, Springer-Verlag, **1434**, 143.

Witkin, A., Kass, M., 1988, Spacetime Constraints, *Computer Graphics*, **22**, 159.

Zeltzer, D., 1991, Task Level Graphical Simulation: Abstraction, Representation, and Control, *Making Them Move*, edited by Badler.

Chapter 4

Virtual Humans on Stage

Nadia Magnenat-Thalmann
Laurent Moccozet
MIRALab, Centre Universitaire d'Informatique
Université de Genève
Nadia.Thalmann@cui.unige.ch
Laurent.Moccozet@cui.unige.ch
http://www.miralab.unige.ch

4.1. Abstract

We describe a framework for modeling human body parts, animating them, and putting them in "virtual" scene in the context of Virtual Reality. Modeling human bodies for Virtual Reality environments defines some constraints on the way body parts have to be modeled. The modeling has to meet two major goals: the resulting body must be able to be animated, which requires more data structures than the 3D geometric shape, and the animation must run in real-time. To fulfill these two objectives, the modeling of the human body is then much more complex than simply reconstructing its 3D shape. Many more aspects have to be considered, which this paper addresses by describing some efficient approaches. We consider a virtual human as a set of two types of data: the 3D shape and the animation structures. Both are taken into consideration in the modeling process. The resulting human body model can be fed to a scalable human body animation model to simulate virtual humans and clones ready to evolve inside a virtual environment. The resulting functional virtual humans have then to be put in scene in order to move inside their virtual environment and interact with it. Various case studies are described including standalone and distributed applications. Realistic virtual humans should not be limited to simulating their body. A virtual human should be considered as a whole with his accessories. Among these accessories, the most important are the clothes. We open the way towards dressed virtual humans, by defining the current status of our work in terms of interactive clothes design, virtual humans clothing and clothes simulation, and raise some still open issues in order to integrate clothes and virtual dressed humans in VR environments.

4.2. Introduction

Equipment such as 3D scanners able to capture the 3D shape of human bodies seem an appealing way to model human bodies. Although the method seems as simple as taking a photo, the approach is not so straightforward and the resulting human body shape can not directly be used inside a Virtual Reality environment. The manipulation of such a device has various levels of difficulties to obtain acceptable results. For example, defining the appropriate settings of the device is generally not straightforward. The main drawback lies in the post-processing required getting a usable virtual human. In order to get a proper 3D shape, the result requires a geometric modeling post-process, particularly in terms of the amount of geometric data and surface correction, such as smoothing surfaces or filling holes. In a next step, animating the 3D shape inside a Virtual Reality environment implies to combine it with the animation structure. This requires additional information that the scanning process can not provide. We can illustrate the problem with one example: animating the face requires the eyes to be independent from the rest of the head. When the face is built using a 3D scanner, first, only the visible parts of the eyes are modeled, second, they make one single surface with the rest of the face. In this chapter, we describe some approaches for modeling human body parts, particularly human head and hands. We also describe how they can be further animated and focus our attention on human hands. Our approach optimizes the modeling of virtual humans in order to get a "ready to animate" virtual human. The basic idea is to consider a virtual human as a combined set of data, including the 3D shape but also the structure to animate the shape inside a Virtual Environment. The human modeling approach described further starts from default virtual humans templates, including shape and animation structures, and modifies the shape to create a new virtual human. The animation structure is updated according to the changes applied to the 3D shape template. Shape modifications are transmitted to the attached animation structures in order to keep the set of data consistent. Thanks to of the modeling process, we get a "ready to animate" virtual human. Our integrated approach aims also to address the modeling of the data structure needed for the animation.

Defining virtual humans for Virtual Reality applications puts heavy constraints on the model used to represent and animate them. Traditionally, the optimization needed to display static virtual scenes in Virtual Reality applications consists in reducing as much as possible the number of geometric primitives proceeding by the graphic pipeline. This can be done by selecting which parts of the virtual scene are visible, and by using different representations of the same object with different amounts of geometric primitives and switching dynamically between them according to the current state of the scene. The latter technique is known as Level Of Details (LODs). Introducing realistic virtual humans changes the context. First, we introduce an additional process in the application pipeline, where the geometry is modified before being displayed. Then the LOD approach is no more accurate to optimize the scene display. It would require the application of the deformation process to all the available LODs representing a human body, which would slow down the whole process, moreover deforming 3D surfaces requires a minimum density of geometric primitives in order to give reasonable visual results, which greatly reduce the possible optimization with

LODs. It must also be noted that current work on surfaces simplifications are focused on rigid objects. Applying such a simplification algorithm to a virtual body would not be satisfying, as it would not take into account the fact that the currently rigid body will be further deformed, and the shape will change. In the same way we point out that the body modeling step must integrate the animation structure, we can say that the modeling and animation steps must integrate the constraints related to real-time animation.

Once virtual humans are built and ready for animation, they have to be put in scene. It means that they have to be guided and controlled, but also that they have to interact with this environment. This requires an open and versatile framework allowing to control virtual humans with various means, from keyboard to motion capture devices. We provide such an environment with the Virtual Life Network (VLNET). The VLNET core performs all the basic system tasks: networking, rendering, visual data base management, virtual humans management including body movement and facial expressions. It has been used to build many real-time experiments and performances.

The chapter is organized as follows. In a first section we briefly describe the traditional approach used to define virtual articulated characters, available in commercial software. We then describe our human body parts shape modeling approaches, classified according to their user interaction requirements. The following section focuses on the animation models defined for human body parts. As one key aspect of virtual reality is real-time, we pay particular attention to the real-time animation aspect. Thanks to the techniques described in the previous parts, we get "ready to animate" virtual humans. We then describe the virtual environment where Virtual Humans are set in scene: VLNET. Different real-time interactive performances are presented to demonstrate the efficiency of the combination of our virtual humans simulation framework with our virtual "stage" environment. Before concluding, we describe our interactive clothing environment to design clothes, dress virtual humans, simulate clothes, and propose a direction to obtain real-time virtual dressed humans.

4.3. Human Body Modeling: Traditional Approach

Defining a virtual human or clone currently requires 4 steps [Maestri 1996]:

1. Modeling the 3D geometric skin,
2. Modeling the material properties,
3. Mapping the skin onto the skeleton,
4. Attaching the skin to the skeleton.

In most of the existing approaches, these steps are independent from each other. Moreover they imply many human interactions, each of them requires a lot of artistically, anatomical and technical skills. This is illustrated by the traditional approach available in current commercial animation software such as Alias PowerAnimator [Endrody 1998], Softimage [Wilcox 1997] or 3DS MAX Forcade 1996][Apostoliades 1998]. The designer needs first to create the 3D geometric skin, this can be made either by modeling it, or by importing it. The next step requires building the 3D skeleton, then to interactively place the 3D skin around the skeleton. This step is very tedious, as each joint of the skeleton has to be correctly placed inside the geometric skin. It is important to notice that the quality of the further skin deformations are closely linked to this step: if the skeleton is not correctly placed relative to the skin, whatever animation technique is used, the result will not be at all satisfying. Once the skeleton is placed inside the skin, the next operation consists of attaching each vertex of the skin to a segment of the skeleton. The character is then ready to be animated, although there is an additional step where the designer has to test all degrees of freedom of the skeleton, check how the skin is deformed and eventually interactively correct the attachment of the skin with regards to the skeleton segments. The skin-skeleton linking step is quite tedious, particularly if the body shape is modeled independently from the way it will be animated.

The previous approach is quite fine as long as we want to model one "unique" virtual human, such as the famous Kyoko Date for example, but any new virtual human requires to start almost from scratch. On the other hand, virtual humans are becoming more and more popular in Virtual Environments, such as virtual clones, or virtual representation (avatar) of users. This requires to be able to model a ready for animation virtual body in a very short time, and to be able to reproduce many with very few manipulations.

Another commercially available approach to modeling virtual humans is proposed in virtual environments designed for human factor studies. Usually, the shape of these humans is rather crude, and made of a set of rigid unconnected 3D pieces. The main advantage of method is that any virtual humans can be modelled using some anthropometric parameters. For example, in software like SafeWork [Safework], 103 anthropometric variables are available to parameterize the human body. In Mannequin-Pro, a anthropometric database is provided. The modeling step is then much easier, and the resulting virtual body is ready for animation, but the visual aspect of the human body is not very sophisticated. This visual aspect is outside of the scope of such software, because they care about the correctness of the body proportions for ergonomic analysis. Such an approach for modeling human body is simple as long as the geometric representation used to model the body has a low geometric primitives density, and is made of a set of rigid segments corresponding to the body limbs. With such a representation, scaling body parts only consists of scaling the corresponding rigid skin segment along the axis of the limb. Such an approach applied a single sophisticated body surface is not so straightforward and requires complicated methods. It must work so that, once the the default body is scaled in order to fit some given anthropometric parameters, the resulting body keeps the same visual quality as the default one.

We describe in the next section a framework for modeling human body that integrates the animation structure in the shape modeling process, and aim to reduce as much as possible the interactive processes in the body modeling step.

4.4. Modeling Body Parts

The first step to perform, in order to simulate a virtual human, consists of modeling the shape of the body. Many approaches can be considered, but it has to be kept in mind that the shape modeling step has to be consistent with the way the body will be animated. Although 3D shape capturing equipment is getting more and more popular, allowing anybody to have a 3D copy of their face or body, this is far from being enough to obtain a clone. 3D scanners are only able to extract the shape of body parts or a complete body in a given configuration. It is not able to extract the information related to the way the body will move and deform. All the information related to the dynamic aspect of the body have to be empirically added in a post-process. Such as post-process is tedious and slow, and it is not possible to rely on the success of the result before trying it. This results most of the time in a loop of trials and corrections before getting a satisfying result.

We present three approaches sorted by the level of interactivity needed to get the final body part shape. The first one is fully interactive and consists of sculpting new body parts by sculpting template body parts. The approach is illustrated with hands and faces. In order to create virtual clones, a picture of the person to be cloned can be used as a guideline in the sculpting process. The second approach is semi-automatic. The user has to extract or define the main features of the body part to be modelled. The features are further used to automatically model the complete body part. The last method is fully automatic. Body parts can be parameterized, based on morphological differences between people (big, fat, thin, etc). New body parts can be modeled from a template by providing a set of values corresponding to the parameters that control the shape of body parts. In any of these approaches, the modeling starts from a given template, involving the geometric shape and the animation structure. The shape of a new body part can be created interactively, semi-automatically or automatically by deforming the template shape. The modifications of the template shape are then sent to the animation structure so that the resulting new body parts is ready to be further animated without additional manipulations or interactions.

4.4.1. Surface Sculpting Body Modeling

Our sculpting approach [Paouri 1991] for modeling body parts is based on an interactive sculpting tool for creating and modifying an irregular triangular meshed surface. The tool permits operations like adding, deleting, modifying, assembling and splitting primitives of the surface. The methodology employed for interaction corresponds to the traditional sculpting with clay [LeBlanc 1991]. Utilities for local and global deformations are provided to alter the shape attributes of the surface. This allows the creation of an object starting from a given object by transforming it into another one. This approach is often used, as a designer normally likes to start with a simpler shape primitive (sphere, cylinder, etc.) and then obtains the final model through successive transformations. Combining topological operators, such as triangle splitting together with deformation operators, such as bending provide an efficient tool for modeling objects. This tool is used primarily for transforming the already existing models of some body parts to the desired shapes [Magnenat Thalmann 1995]

For example, the existing surface model of hands was refined to give a feminine look. Operations like assigning colors to the nails of the hands and stretching them have also been performed. Figure 1 shows the modeling of hands starting from an initial existing model.

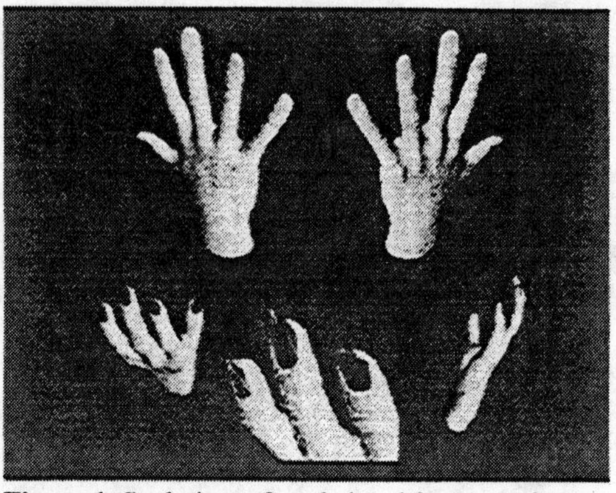

Figure 1. Sculpting a female hand from a male one.

Using this *Sculptor* program, instead of a scanning device, allows us to take into account the constraints of real-time animation while modeling the head. Real-time manipulation of objects requires as fewer polygons as possible, while more polygons are required for the beautification and animation of the 3D shape. Designing the head with this program allows us to directly create an object with the appropriate number of polygons. This can be done knowing which region has to be animated (this requires a lot of polygons) and which requires less, or no animation at all (this requires as few polygons as possible). Designers are also able to model simpler objects knowing the texture will add specific details, like wrinkles and shadows, to the object. Although many automatic mesh simplification techniques are available [Turk 1992][Schroeder 1992][Hoppe 1996][Garland 1997], they focus on rigid objects and do not take into account the further deformations that will be applied to the geometric surface. Starting from a template head accelerates the creation process. shows the modeling of a face starting from an already existing one, the adaptive template. The resulting face includes both the geometric shape and the animation structure required for further animation [Kalra 1991].

The modeling process can also be improved by using pictures of the real face to be modeled as a guideline, such as in Figure 2. For example, we use a specific tool, called a "cutting plane". The plane, where the picture of the face is drawn, can be interactively moved along the main axis of the 3D face. The intersection points between the cutting plane and the surface are drawn and can be selected and manipulated, such as shown in Figure 3. The designer can then interactively adapt the section to match some features of the picture. The regions defining the facial muscles are updated according to the 3D shape modifications, as shown in Figure 4.

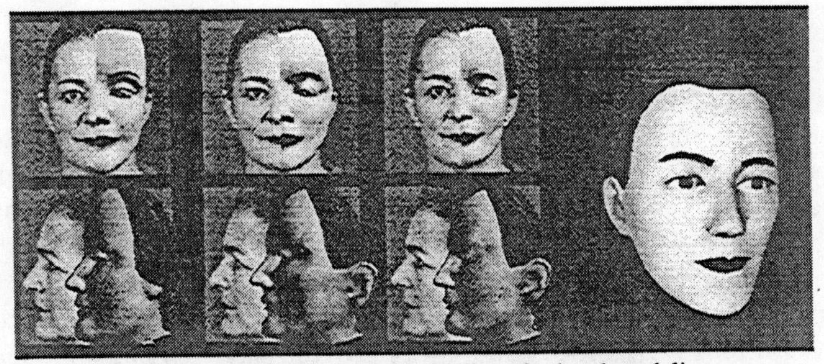

Figure 2. Using pictures as guidelines for head modeling.

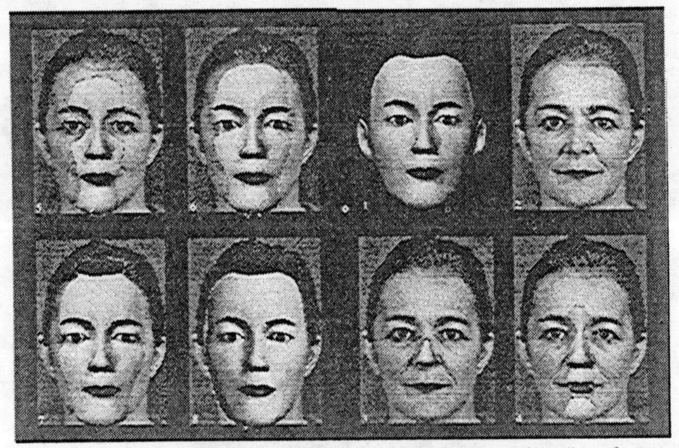

Figure 3. Face modeling with the "cutting plane" tool.

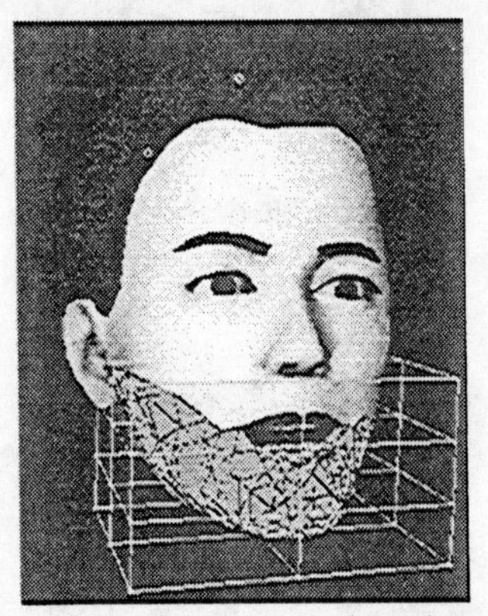

Figure 4. Final face, ready for animation.

4.4.2. Semi-automatic Body Modeling

Our approach [Lee 1997] to reconstruct a real face from two orthogonal views is particularly suited for face cloning. In this reconstruction, only some points (so called feature points) are extracted. These are the most characteristic points, needed to recognize people, that can be detected from front and side views. It is a simple approach and does not require high cost equipment. The extracted features are then used to constrain the deformation of a template 3D face. The deformation model is further described to simulate the muscular layer in the hand animation model, and called DFFD [Moccozet 1997]. The overall flow for the reconstruction of 3D head is shown in Figure 5. The method can also automatically fit the texture image onto the modified 3D model.

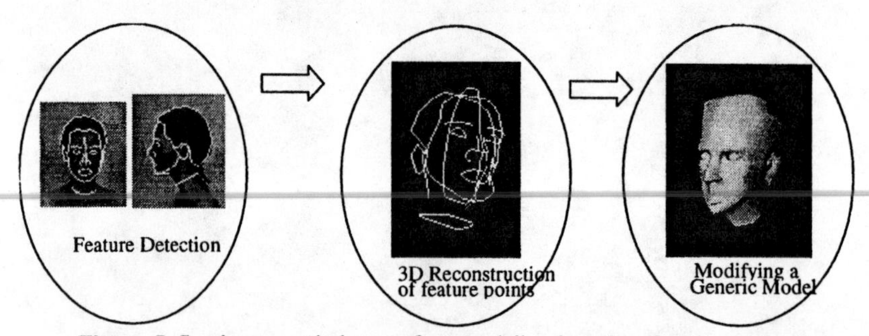

Figure 5. Semi-automatic human face modeling based on features extraction.

4.4.3. Automatic Parameterized Body Modeling

The proposed hand simulation model, further described in this paper, allows the modeling of morphological variations. Deformations of the muscular layer can be parameterized according to some morphological changes, such as hand thickness, or skeleton finger length. The muscle layer is first fitted to the skeleton, or scaled to the given morphological parameters. The muscular deformation function is then applied to the hand surface to create a new hand. The resulting new hands can be directly used for animation.

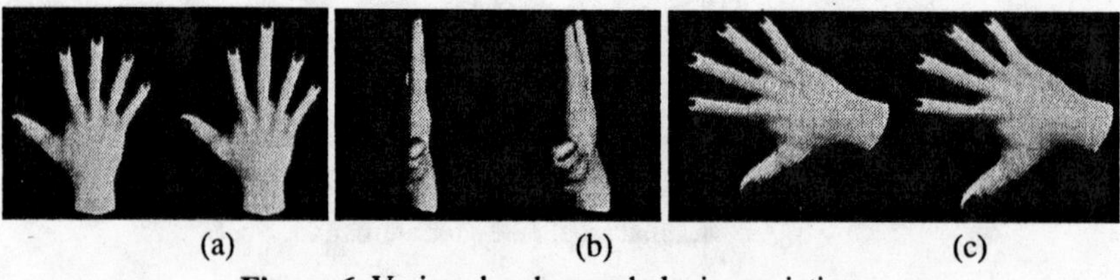

(a) (b) (c)

Figure 6. Various hands morphologies variations.

Figure 6 shows some of the possible morphological variations of hands. Figure 6.a and Figure 6.b show local modification changes: in 6.a, the length of the middle finger is modified, and in 6.b, the thumb is spread from the rest of the hand and its size is modified. These two morphology changes are parameterized by the changes of the underlying skeleton. In Figure 6.c, a global morphology change is shown: the thickness of the hand is increased. Global changes are parameterized independently from the skeleton, as it does not result from a skeleton modification.

4.4.4. Texture Fitting

For virtual humans, the texture can add a grain to the skin, including the color details like color variation for the hair and mouth. These features require correlation between the image and the 3D object. A simple projection is not always sufficient to realize this correlation: the object, when designed by hand, can be slightly different from the real image. Therefore an interactive fitting of the texture is required [Litwinowicz 1994]. In Figure 7, the wrinkles of the hands have been fitted to the morphology of our 3D model. Figure 7.a shows the texture, which has been applied, to the 3D hand shown in Figure 7.b.

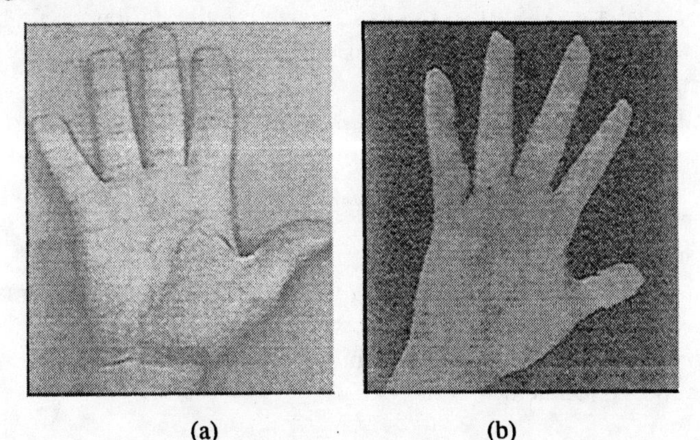

(a) (b)

Figure 7. Texture applied on a 3D hand.

Our approach for fitting the texture to the features of the 3D object [Sannier 1997] enables the designer to interactively select a few 3D points on the object. These 3D points are then projected onto the 2D image. The projection can be chosen and set interactively. With this method, we obtain the texture coordinates for the selected 3D points. This approach allows the storage of the texture data in the body part templates. As new body parts are modeled by deforming the shape of the template, the texture information remains consistent for the new body parts.

4.5. Body Parts Animation

4.5.1. Hands Animation

Hands have had a very specific treatment in real-time human simulations: it is generally considered too expensive to use a deformable hand model according to the little advantages it brings to the global visual result. This approach is closely linked to the optimization approach generally used in virtual environment: the Level of Detail

(LOD) [Funkhauser 1993]. In this approach, the importance of an object is mainly linked to its rendered size relative to the size of the final picture. According to this hypothesis, hands should not be given a lot of attention. We want first to justify the need for providing a real-time and accurate hand simulation model by first briefly underlining the importance of hands in human simulation inside virtual environments. We will then show how we developed a dedicated simulation model for the hand, and then parameterized it in order to perform realistic hand simulation for real-time environment.

4.5.1.1. Hands in Human Simulation

Hands represent a very small part of the whole body. If we refer to artists' rules for human drawing, the ideal proportions are defined so that the whole body is 8 times the height of the head, and the height of the hand is equal to the height of the head. The width of the hand is equal to half of its height. These are canonical measurements, but they give a good idea of the hand's size relative to the whole body. This relative size aspect has to be taken into consideration, but the importance of hand is not restricted to this aspect.

According to A. Mulder [Mulder 1996] hand gestures can be classified into three main categories, which are described as follows:

- *Semiotic*: to communicate meaningful information and results from shared cultural experience.
- *Ergotic*: is associated with the notion of work and the capacity of humans to manipulate the physical world, create artifacts.
- *Epistemic*: allows humans to learn from the environment through tactile experience or hepatic exploration.

These three categories show how important the hand is in simulating a realistic human when interacting with a virtual environment. The hand is both an effect and a sensor. It is a gateway between humans and their environment. This implies that hands are a center of interest, and that despite their size, many situations during the simulation will focus on them. In order to provide a convincing visual representation, they have to be appropriately modeled and simulated.

Hands concentrate a great number of Degrees of Freedom. Our skeleton model counts around 120 total DOF. Each hand contains 25 DOF, which means that around 40 percent of the DOF. Therefore, the hands are the most flexible part of the body, and the total number of possible postures and movements is very large. Thus, the hand may have a high level of deformation, concentrated in a very short area. Moreover, a brief look at anyone's hands shows that in addition to muscular action, the skin deformation is controlled by the main hand lines associated to the joints of the skeleton.

Although this size aspect is important, we have shown that the importance of the hand, as part of a human simulation, requires more attention than has previously been devoted to it: a set of rigid articulated skin pieces. Although the hand is a part of the body, we have shown that its particularities, such as the importance of hand lines regarding the skin deformation requires a dedicated model. We propose a model for hand simulation suited for real-time to be used in conjunction with the traditional rigid skin piece approach.

4.5.1.2. Basic Hand Multilayer Model

We have proposed a multilayer model for human hands simulation [Moccozet 1997], as a part of the HUMANOID environment [Boulic 1995]. The basic model, following the traditional multilayer model for articulated deformable character, is subdivided into three structural layers (skeleton, muscle and skin). We combine the approaches of Chadwick et al. [Chadwick 1989] and Delingette et al. [Delingette 1993] to design the three layers structure deformation model for hand animation. The intermediate muscular layer, that maps joint angle variations of the basic skeleton layer into geometric skin deformation operations, is based on a generalized Free-Form Deformations (FFD's) [Sederberg 1986] model called Dirichlet Free-Form Deformations (DFFD's) [Moccozet 1997]. The structural muscle layer is modeled by a set of control points attached to the skeleton, that approximate the shape of the hand's skin. The motion of the skeleton controls the displacements of the control points. Once the control point set is fitted to the current configuration of the skeleton, the generalized FFD function is applied to the triangle mesh that represents the geometric skin. We now briefly describe the main features of the basic geometric deformable model involved in the multilayer hand model, and its application to simulate muscular action on the skin. The main objective of the resulting model is to simulate how the muscles deform the skin and how hand lines control the skin deformations.

The relationship between the control points and the object to deform is based on a local coordinate system called Natural Neighbors (NN) or Sibson coordinate system [Sibson 1980][Farin 1990][Sambridge 1995]. The geometric deformation model is thoroughly described in "Hands Modeling and Animation for Virtual Humans" [[Moccozet 1996]. This local coordinate system allows, for each vertex of the surface, the automatic definition of a subset of control points whose displacements will affect it. This subset is used to build a deformation function similar to the one defined for FFDs'. The deformation function is defined inside the convex hull of the control points. It interpolates the displacements of the control points to the vertices of the deformed surface. Among all the properties of the resulting model, it must be noted that there is no constraint on the location of the control points, or on the shape of their convex hull. Moreover, there is no need to explicitly define the topology of the control points set. All of the FFD extensions can be applied to DFFD. Among them are two of particular interests to our hand simulation model. Weights can be assigned to control points and define Rational DFFD (RFFD) [Kalra 1992] with an additional degree of freedom to control deformations. Direct surface manipulation [Hsu 1992] can also be performed with the basic model without having to use any estimation method. As any location in the 3D space can be defined as a control point, assigning a control point to a vertex on the surface allows us to directly control its location: any displacement applied to the constraint control point is integrally transmitted to the associated vertex of the surface. Thanks to the flexibility introduced by our generalized FFD model, we provide an efficient tool to design an intermediate layer to interface the skeleton motion and the skin deformation, by simulating the muscles and hands lines behavior.

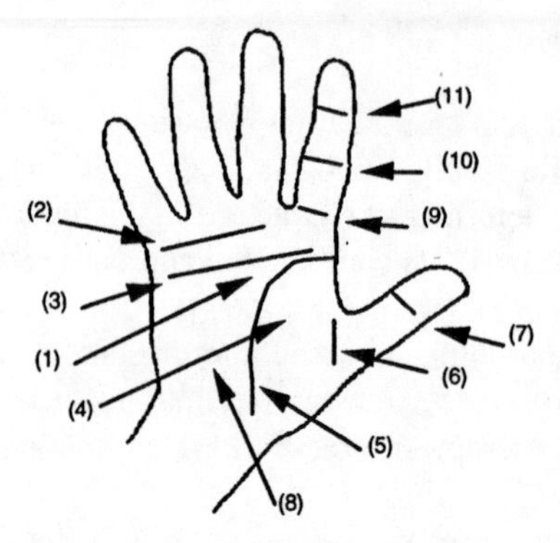

Figure 8. Topography of the hand.

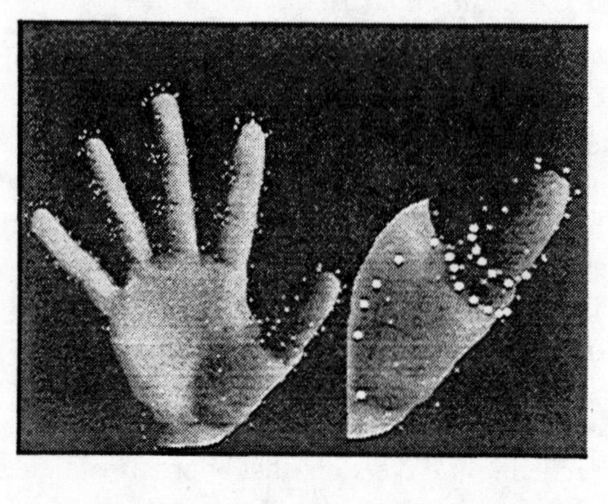

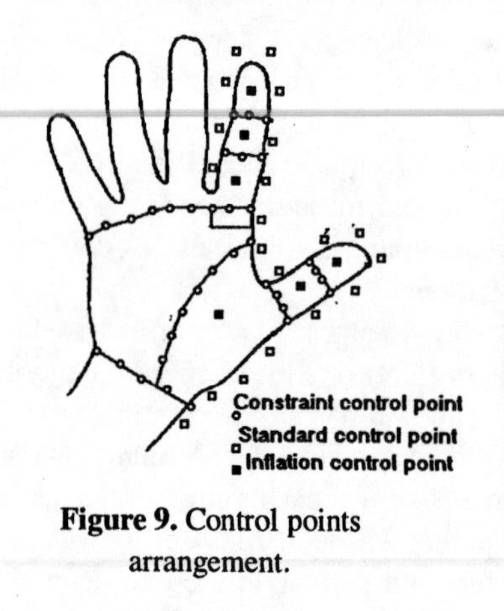

Figure 9. Control points arrangement.

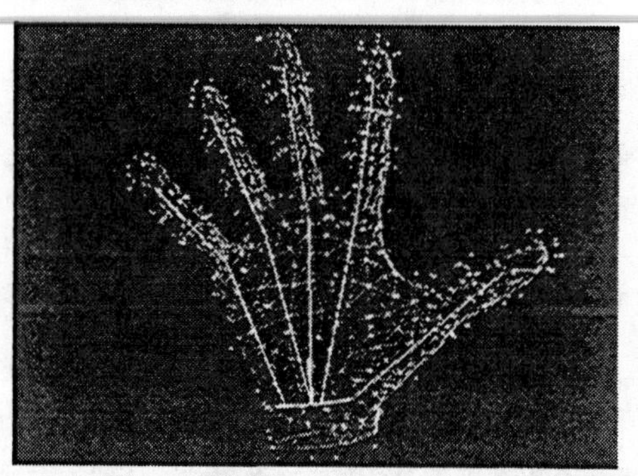

Figure 10. Control points set and wrinkles design.

The set of control points is built in order to match a simplified surface hand topography. From the observation of a hand, especially its topography [Kapandji 1980], we derive the following basic data structure for our hand model, that we call a wrinkle, as shown in Figure 9:

- The wrinkle itself, which is a set of constraint control points. They are generally selected around the joint and form a closed 3D line. We call such points wrinkle control points.
- Two points among the wrinkle control points. They are used to define the axis on which the associated skeleton joint should lie. In this way, a skeleton model can be easily adapted to the hand's skin. This data allows an easy and realistic hand-skeleton mapping by defining an implicit skeleton to which the skeleton can be fitted.

- A mixed set of control points and constraint control points. They surround the upper part of the hand surface that will be affected by the rotation of the joint associated with the current wrinkle. We call these points influenced wrinkle control points, as they are influenced by the rotation of the wrinkle itself.
- One control point, called an inflation control point, which will be used to simulate inflation by the upper limb associated with the joint.

For each wrinkle, the muscle layer gets the joint angle variation from the skeleton layer. If the rotation angle is α, the wrinkle itself is rotated at an angle of $\alpha/2$, and the set of influenced control points is rotated at α. At the rest position, all control points have a weight of 1. When the joint angles vary, the weights of the inflation control points vary accordingly. This point is placed on the mesh so that when its weight increases, it attracts the mesh, simulating the skin inflation due to muscle contraction.

Figure 8 shows the simplified surface hand topography we want to model with the important hand lines associated with the underlying skeleton joints [(1) palm; (2) upper transversal line; (3) lower transversal line; (4) thenar eminence; (5) thenar line; (6) thumb first line; (7) thumb second line; (8) hypothenar eminence; (9) finger first line; (10) finger second line; (11) finger third line]. Figure 10 shows how control points, constraint control points and inflation control points are designed around the surface of the hand to build the control points set and the different wrinkles.

4.5.1.3. Optimization Approach

The complexity of our basic muscular deformation model is linked to:

- the degree of the deformation function,
- the number of control points involved.

Our goal in this optimization step is to provide a deformation function that is able to work at different levels, and introduce a Deformation Level of Detail, in a similar way as Geometric Level of Detail. The optimization is achieved by parameterizing the deformation function with two features that constrain the function complexity. The real-time hand simulation model is working on a fixed Geometric LOD. This is motivated by the fact that a realistic hand shapes requires a high resolution, and that performing deformations is only worthwhile for a minimum resolution of the deformed surface.

As for basic FFDs, the deformation function is a cubic function of the local coordinates. The properties of the deformation function make it possible to use it at lower degrees with a minimal loss in the continuity of the deformation. We can then choose between a linear, a quadratic or a cubic deformation function.

The total number of control points involved in the deformation of a vertex is not predefined and depends on the local configuration of the set of control points at the location of the vertex inside the convex hull of control points. The number of control points can be controlled and constrained between four control points up to the "natural" number of NN control points. It must be noted that limiting the number of control points to 4 results in a continuity problem. This has to be considered as the price to pay for the gain in speed. Figure 10 shows, for the same hand posture, 3 different results with various constrained Sibson control points. Figure 11.a shows the basic DFFD function, whereas Sibson control points are constrained to be a maximum of 9 in Figure 11.b and 4 in Figure 11.c. The hand's figure contains around 1500 vertices

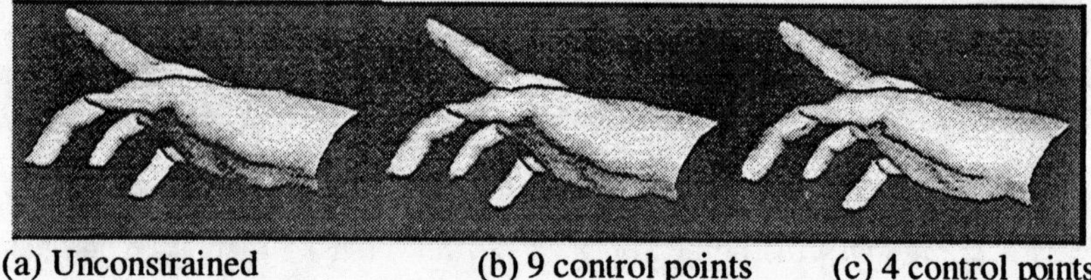

(a) Unconstrained (b) 9 control points (c) 4 control points

Figure 11. Hand posture with various constrained Sibson control points.

As a conclusion to the deformation function optimization, we have shown that the complexity of the deformation function can be controlled and parameterized according to the degree of the function and the maximum number of control points used. If the deformation function name is DFFD, Deformation LODs can be defined by setting the two parameters Degree, and MaxControl of the DFFD function: DFFD (Degree, MaxControl). Degree can take its value between 1, 2 and 3, and MaxControl takes value higher than 4.

4.5.2. Facial Animation

In our real-time human animation system, a face is considered as a separate entity from the rest of the body. This is due to its particular animation requirements. The face unlike the body is not based on a skeleton. We employ a different approach from body animation for the deformation and animation of a face, based on pseudo muscle design.

Developing a facial model requires a framework for describing geometric shapes and animation capabilities. Attributes such as surface color and textures must also be taken into account. Static models are inadequate for our purposes; the model must allow for animation. The way facial geometry is modeled is motivated largely by its animation potential as considered in our system. Facial animation requires a deformation controller or a model for deforming the facial geometry [Kalra 1992]. In addition, a high level specification of facial motion is used for controlling the movements [Kalra 1991].

4.5.2.1. Facial Deformation Model

In our facial model the skin surface of a human face, being an irregular structure, is considered as a polygonal mesh. Muscular activity is simulated using Rational Free Form Deformations (RFFD) [Kalra 1992]. To simulate the effects of muscle actions on the skin of virtual human face, we define regions on the mesh corresponding to the anatomical descriptions of the regions where a muscle is desired. For example, regions are defined for eyebrows, cheeks, mouth, jaw, eyes, etc. A control lattice is then defined on the region of interest. Muscle actions to stretch, expand, and compress the inside geometry of the ace are simulated by displacing or changing the weight of the control points. The region inside the control lattice deforms like a flexible volume according to the displacement and weight of each control point. A stiffness factor is specified for each point, which controls the amount of deformation that can be applied to each point; a high stiffness factor means less deformation is permissible. This deformation model for simulating muscle is simple and easy to perform, natural and intuitive to apply and efficient to use for real-time applications.

4.5.2.2. Facial Motion Control

Specification and animation of facial animation muscle actions may be a tedious task. There is a definite need for higher level specification that would avoid setting up the parameters involved for muscular actions when producing an animation sequence. The Facial Action Coding System (FACS) [Ekman 1978] has been used extensively to provide a higher level specification when generating facial expressions, particularly in a nonverbal communication context.

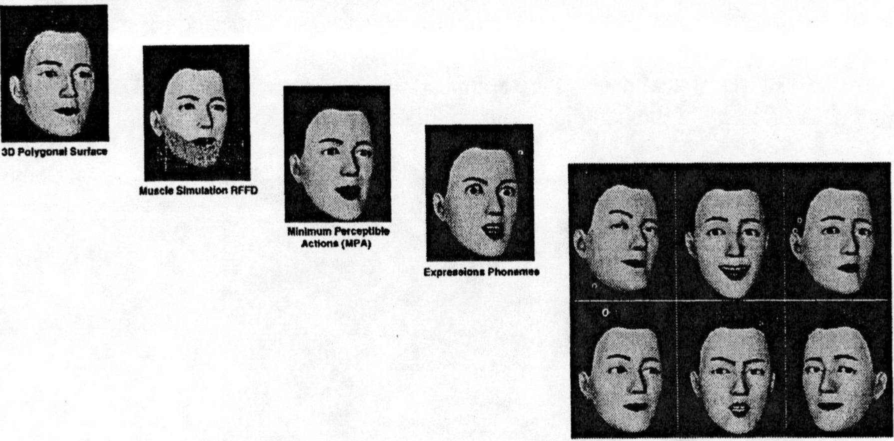

Figure 12. Different levels of facial motion control.

In our multi-level approach (cf. Figure 12), we define basic motion parameters as Minimum Perceptible Action (MPA). Each MPA has a corresponding set of visible features such as movement of eyebrows, jaw, or mouth and others occurring as a result of muscle contractions. The MPAs are used for defining both the facial expressions and the visemes[1]. There are 65 MPAs, such as open_mouth, close_upper_eyelids or raise_corner_lip) used in our system which allow us to construct practically any expression and viseme. At the highest level, animation is con-

[1] Visemes are defined as the animated shape of the face resulting from the motion of the mouth (lips and jaw) and corresponding to one or more phonemes (different phonemes may correspond to the same viseme).

trolled by a script containing speech and emotions with their duration. Depending on the type of application and input, different levels of animation control can be utilized. There are three different input methods used for our real facial time animation module: video, audio or speech, and predefined actions.

4.5.3. Body Deformation

An envelope of the virtual human is created using BodyBuilder [Boulic 1995] The envelope is defined by a set of spline surfaces. The underlying skeleton is a hierarchical model comprising 32 joints corresponding to 74 DOFs (Degrees Of Freedom). The final step is to create the cross-sectional data for the body. Indeed, human limbs and truck exhibit a cylindrical topology, therefore, a natural centric axis is inherent for each body part. Scanning each body part along this axis allows the creation of cross sections at regular intervals. All sections of a limb contain the same number of points; it is therefore possible to reconstruct the triangle mesh for each body part directly from the contour points. The mesh is updated automatically by simply rotating the contour. The complete algorithm developed to move the contour is explained in "Fast Realistic Human Body deformations for Animation and VR Applications" [Shen 1996], but the main idea is to use the angle between two connected segments in the skeleton. In

Figure 13 three virtual dancers are animated using this methodology. The more the contour is far away of the body joint (in real distance), the less the contour is moved.

Figure 13. Three animated bodies.

4.6. Putting Virtual Humans on Virtual Stage

4.6.1. Virtual Stage Environment

Virtual Life Network (VLNET) is a general purpose client/server Networked Collaborative Virtual Environment (NCVE) system [Carlsson 1993][Macedonia 1994] providing an efficient framework to integrate and control highly realistic virtual humans [Pandzic 1997a]. VLNET achieves great versatility through its open architecture with a set of interfaces allowing external applications to control the system functionality.

Figure 14 presents a simplified overview of the architecture of a VLNET client. The VLNET core performs all the basic system tasks: networking, rendering, visual data base management, virtual humans management including body movement and facial expressions. A set of simple shared memory interfaces is provided through which external applications can control VLNET. These interfaces are also used by the VLNET drivers. The drivers are small service applications provided as part of VLNET system that can be used to solve some standard tasks, e.g. generate walking motion, support navigation devices like mouse, SpaceBall, Flock of Birds etc. The connection of drivers and external applications to VLNET is established dynamically at runtime based on the VLNET command line.

- *The Facial Expression Interface* is used to control expressions of the user's face. The expressions are defined using the Minimal Perceptible Actions (MPAs) [Kalra 1993]. The MPAs provide a complete set of basic facial actions. By using them it is possible to define any facial expression.
- *The Body Posture Interface* controls the motion of the user's body. The postures are defined using a set of joint angles defining the 72 degrees of freedom of the skeleton model [Boulic 1995] used in VLNET.
- *The Navigation Interface* is used for navigation, hand and head movement, basic object manipulation and basic system control. All movements are expressed using matrices. The basic manipulation includes picking objects up, carrying them and letting them go, as well as grouping and ungrouping of objects. The system control provides access to some system functions that are usually accessed by keystrokes, e.g. changing drawing modes, toggling texturing, displaying statistics.
- *The Object Behavior Interface* is used to control the behavior of objects. Currently it is limited to the controlling of motion and scaling, defined by matrices passed to the interface. It is also used to handle the sound objects, i.e. objects that have prerecorded sounds attached to them. The Object Behavior Interface can be used to trigger these sounds.
- *The Video Interface* is used to stream video texture (as well as static textures) onto any object in the environment. The Alpha channel can be used for blending and achieving effects of mixing real and virtual objects/persons. The interface accepts requests containing the image (bitmap) and the ID of an object on which the image is to be mapped. The image is distributed and mapped on the requested object at all sites.
- *The Text Interface* is used to send and receive text messages to and from other users. An inquiry can be made through the text interface to check if there are any messages, and the messages can be read. The interface gives the ID of the sender for each received message. A message sent through the text interface is passed to all other users in a VLNET session.
- *The Information Interface* is used by external applications to gather information about the environment from VLNET. It provides high-level information while isolating the external application from the VLNET implementation details. It also allows two ways of obtaining information, namely the request-and-reply mechanism and the event mechanism.

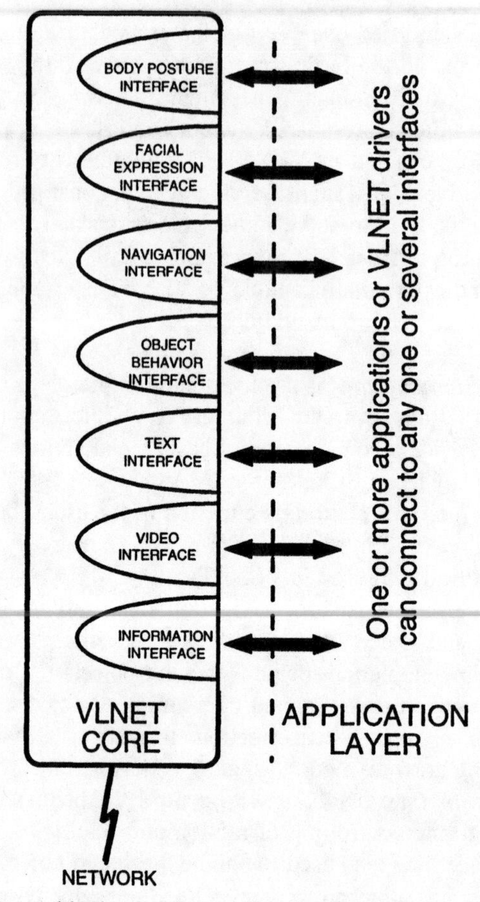

Figure 14. Simplified view of VLNET client architecture.

Figure 15 illustrates the information flow between the program controlling the autonomous actors and the VLNET system. The following information can be requested from VLNET through the Information Interface:

- Description (name) of an object in the virtual world, given the object ID,
- List of objects (object IDs and corresponding descriptions) whose description contains the given keywords,
- List of users (user IDs and corresponding descriptions) whose description contains the given keywords,
- Transformation matrix of an object in world coordinates, given the object ID,
- Transformation matrix of a user in world coordinates, given the user ID,
- Number of objects in the virtual world that is picked by an user, given the user ID,
- Number of users in the virtual world,
- Collisions between users/objects.

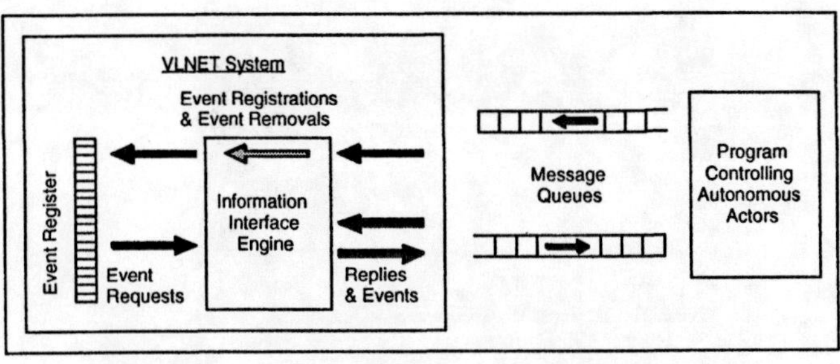

Figure 15. Data flow through the Information Interface.

We have given a brief description of VLNET that allows us to show how external programs can be interfaced to the VLNET system. The focus of this presentation was on VLNET interfaces. The next section describes various live performances in different contexts, demonstrating the versatility of VLNET, where one or more virtual humans can be set in scene and interact with their environment and with each others. For more details on VLNET the reader is directed to references [Capin 1997][Pandzic 1997b].

4.6.2. Some Case Studies

4.6.2.1. CyberDance

CyberDance [Carion 1998] is a new live performance providing interaction between real professional dancers on stage and virtual ones in a computer generated world. This performance uses our latest development in virtual human representation (real-time deformation) together with latest equipment in virtual reality (for motion tracking). Virtual humans were added in the virtual world and one real dancer was tracked to animate his virtual clone in real-time, represented by a fantasy robot.

Figure 16. Motion capture for real-time dancing animation of a virtual clone.

Figure 17. Dancer and his cyber-clone.

Figure 18. Snapshots from the Cyberdance live performance.

Figure 16 and Figure 17 show snapshots of the performance live using motion capture. We can see the dancer being tracked on stage, while the result of this tracking was used to move the virtual robot and displayed in real-time. The audience were able to see both the real and the virtual dancer at the same time. The DataGlove (from Virtual Technologies) and the Flock of Bird (from Ascension Technology Corp.) transmit inputs to the system for hand/body animation. They give absolute data and does not need online-calibration. The data is then used to control the motion of the hand and body skeleton. Figure 18 shows others aspects of the live performance where 3 virtual humans are dancing in different environments.

4.6.2.2. CyberTennis

At the opening session of Telecom Interactive '97 in Geneva, Switzerland, we presented in real-time a virtual, networked, interactive tennis game simulation [Aubel 1998]. This demonstration was a big challenge because, for the first time, we had to put together several different computer related technologies and corresponding software. This had to work in real-time at a specific moment on the exposition site with non-permanent installations. In this demonstration the interactive players were merged into the virtual environment by head mounted displays as shown in Figure 19.b, magnetic flock of bird sensors and data gloves. The University of Geneva player was "live" on stage at the opening session (cf. Figure 19.a) and the other player in the Computer Graphics Lab of EPFL at Lausanne (cf. Figure 19.c), separated by a distance of approximately 60 km.

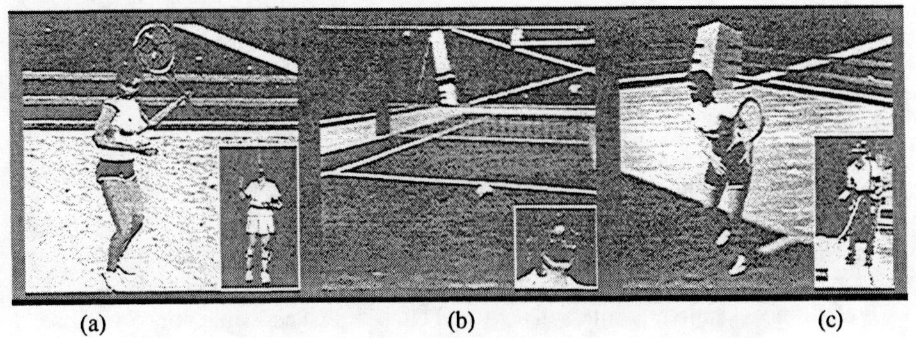

(a) (b) (c)

Figure 19. The virtual players animated by real-time motion capture.

4.6.2.3. Virtual Teleshopping

In collaboration with Chopard, a watch company in Geneva, a teleshopping application was successfully demonstrated between Geneva and Singapore [Pandzic 1997b]. The users were able to communicate with each other within a rich virtual environment representing a watch gallery with several watches exposed. They could examine the watches together, exchange bracelets, and finally choose a watch, such as shown in in Figure 20.

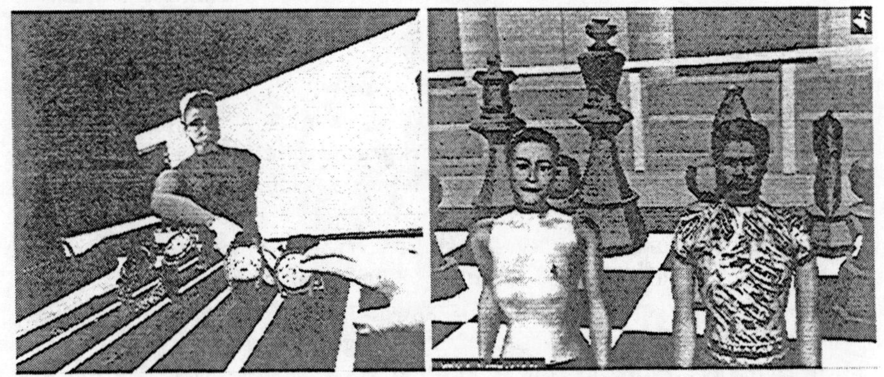

Figure 20. Virtual teleshopping. **Figure 21.** Ready to play chess.

4.6.2.4. Virtual Chess Game

NCVE systems lend themselves to development of all sorts of multi user games. We had successful demonstrations of chess (cf. Figure 21) and other games played between Switzerland and Singapore, as well as between Switzerland and several European countries.

4.7. Towards Real-time Virtual Dressed Humans

4.7.1. Clothes Simulation

Currently, clothes simulation is limited to texture mapping inside Virtual Environments. Of course, it limits the kind of clothes that can be used, moreover the textured clothes are deforming according to the skin surface of the body. It does not reproduce the behavior of the clothes. For example, not wrinkles appear. In a step towards unifying cloth simulation to the wonderful universe of Virtual Reality and dreaming about a world where virtual humans could manipulate cloth in real-time and in a way that seems so natural for us, real humans, we present a contribution for a fast and robust cloth model suited for interactive virtual cloth design and simulation system.

Literature now brings us several techniques for cloth simulation. Many of them present physically based models for simulating in a realistic way fabric pieces based on elastic deformation and collision response. The first of them used simple mechanically-based models, such as relaxation schemes, for simulating objects such as flags or curtains [Weil 1986][Haumann 1988]. More general elastic models were developed for simulating a wide range of deformable objects, including cloth [Terzopoulos 1987, 1988]. Recently, several particle system based models attempted to simulate simple cloth object realistically using experimental fabric deformation data [Breen 1994][Eberhardt 1996]. These models claim to be fast and flexible, as opposed to finite element models [Collier 1991][Kang 1995][Eischen 1996], which are very accurate, but slow and complex to use in situations where behavior models are complicated and where collisions create non-linearity and complex boundary conditions, thus not suited for interactive applications.

Dressing a virtual body is a complex application for these models. It involves the ability to design complex garment shapes, as well as a complex simulation system able to detect and to handle multiple collisions generated between the cloth and the body. Our work contributed to the development and evolution of this topic through several contributions [Lafleur 1991][Carignan 1992]. More recently, we studied how to consider cloth as being an object that could be considered independently from the body which wears it, involving the issues of folding, wrinkling and crumpling, with all the associated problems related to collision detection and response [Volino 1995]. Our work was materialized by several garment design and dressing systems for animated virtual actors [Werner 1993][Volino 1996].

On the other hand, new V.R. technologies and efficient hardware open a very attractive perspective for developing interactive systems where virtual actors would interact autonomously with mechanically animated objects, such as the garment they themselves wear. In a nearer goal, we could take advantage of these new tools for interactively designing garments and dressing virtual actors in ways that are much more natural and close to the "real" way of manipulating fabric.

We provide simulation tools to take a step towards the requirements defined above. Of course, the main problems for interactive or real-time mechanical simulation are related to computation speed issues. We should not however trade away design flexibility and mechanical modelisation accuracy that would lead to unrealistic cloth simulation. Thus, we describe here a mechanical model that allows to modelise elastic behavior of cloth surfaces discretized into irregular triangle meshes, and which is not much more complicated to a simple spring-mass modelisation. This approach combines flexibility [Volino 1995] with simulation speeds [Eberhardt 1996][Hutchinson 1996] which are restricted to regular meshes. Furthermore, a suited integration method has been associated to this model to maximize simulation timesteps and computation speeds without trading away mechanical stability, which is ensured in a very robust way, compatible with all the inaccuracies resulting from most tracking devices used in 3D positioning and V.R. devices.

Beside this, a new approach for handling geometrical and kinematical constraints (such as collision effects or "elastics"), generalization of the collision response process [Volino 1995], ensures collision response as well as integration of different manipulation tools in a robust way that does not alter simulation efficiency and thus makes this system efficient for interactive purposes.

We illustrate the achievements brought by these techniques with the help of examples concerning cloth manipulation, dressing and realtime interactive manipulation.

Cloth simulation algorithms have been extensively studied in the state-of-art research. Current applications allow to design simple garments, to fit them onto a virtual body and to make simple animations. The key issue is now to implement such algorithms into efficient and useful systems that will fully take advantage of the new possibilities brought by these techniques.

Several new problematic arise from integrating cloth simulation algorithms in such systems. First, performance and robustness have to be high enough to provide robust systems that will enable a user to generate quality work in a minimum of effort and time. The main algorithms concerning the simulation aspect of cloth animation are the mechanical model, the numerical simulation algorithms, the collision detection algorithms and the garment design process.

4.7.1.1. Mechanical Model

The objective is to build a mechanical modelization of the cloth behavior that combines adequate realism for representing realistic cloth deformation with computation speed requirements compatible with interactive applications. The designed approaches should be general enough to be implemented in general contexts where the cloth is represented by a triangle mesh structure, possibly irregular that can be dynamically updated and transformed.

Several approaches are defined, ranking from full elasticity evaluation within the mesh triangles [Volino 1995] which aims to perform slow and but realistic modellisation of the fabric behavior using linear viscoelasticity, to faster models that use geometrical approximations to derive a simple spring-mass model into a more realistic model simulating realistically the basic elasticity properties of the fabric [Volino 1997a]. The latter approach is successfully applied in interactive garment systems allowing dynamic manipulation of the cloth elements, such as displacements, cutting and seaming.

4.7.1.2. Numerical Simulation

Our systems mainly use fast numerical integration of particle-system representations of the mechanical models. An efficient Runge-Kutta model [Volino 1997b] has been adapted to provide optimal timestep control through numerical error evaluation. This approach provides a virtually unbreakable model, that can cope with high deformations as well as strong mechanical parameters variations without exhibiting numerical instabilities. These features are important for all the interactive applications.

Still, the stiffness of the simulation model is the major performance-limiting problem. The timestep has to be kept small in order to prevent local vibration instabilities to appear, which highly affects simulations with refined meshes and simulations with rigid materials.

4.7.1.3. Collision Detection

Another important and time-limiting aspect of the cloth simulation problem is collision detection. As a cloth and the underlying body are discretized into a huge number of polygons, the problem is to determine efficiently which polygons are in geometrical contact, in order to apply reaction and friction effects.

The problem is solved efficiently by using an adapted hierarchical scheme [Volino 1994] which is specially adapted for dealing with the self-collision problem, and then extended by orientation-correction algorithms for increased robustness [Volino 1995].

4.7.2. Interactive Tools for Fabric Manipulation and Cloth Design

Through Figure 22.a to Figure 22.g, we illustrate the advances brought by our new system and the new potentialities concerning interactive cloth applications.

With our model, interactive cloth manipulation is now possible, as shown by the following example. Here, we illustrate basic manipulations that are performed on a cloth element discretized to about 400 triangles. The display frame rate varies from 8 to 20 frames per second on a 250MHz R4400 SGI Indigo II, which is quite a comfortable speed for motion perception. Most of the time, the numerical approximation correction scheme allows us to perform the simulation with only one iteration per frame. Figure 22 is a sample of the system's interactive possibilities, the user performing actions that affect immediately and continuously the objects.

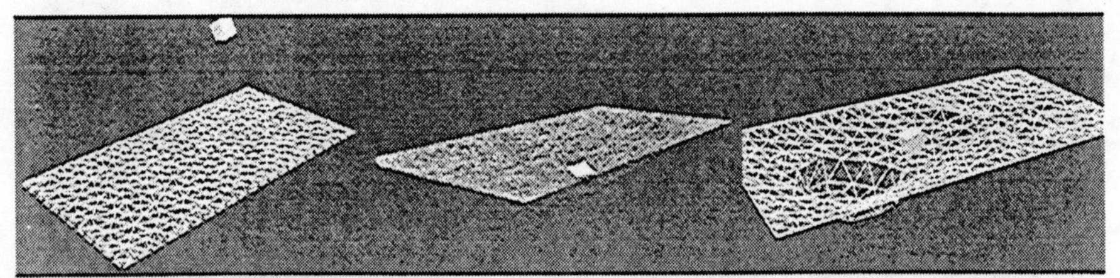

(a) The surface is blocked on its edges, and (b) falls on its own weight. The cube falls, bumps and slips on the surface. (c) The surface may be locally expanded or shrunk.

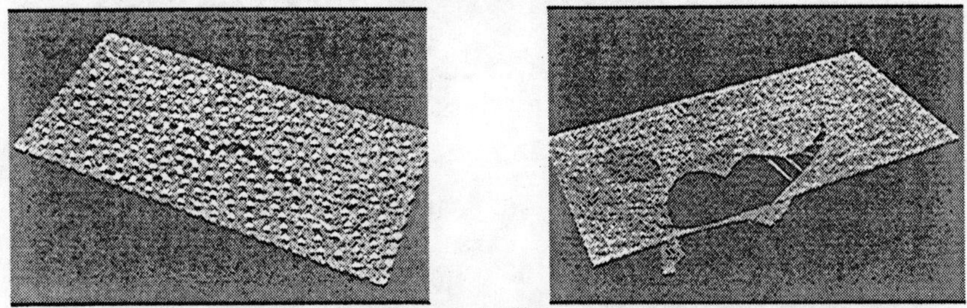

(d) Cuts can then be performed. (e) Extra material can be removed.

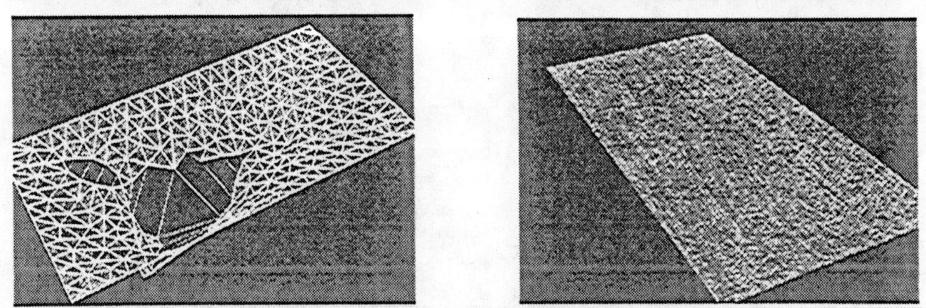

Elastics pull the surface borders together (f) and finally (g) seaming fills the holes.

Figure 22. Interactive clothe manipulation.

4.7.3. Interactive Clothing System

We take advantage of the new advances brought by algorithms in our software [Volino 1995, 1996] by highly improving the garment generation and simulation speed. Figure 23, Figure 24 and Figure 25 illustrate how garments are assembled around the body and seamed together.

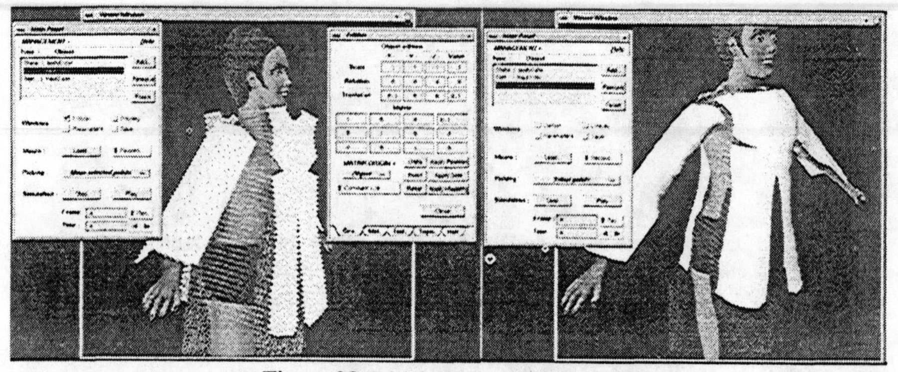

Figure 23. Initial panel and seaming elastics.

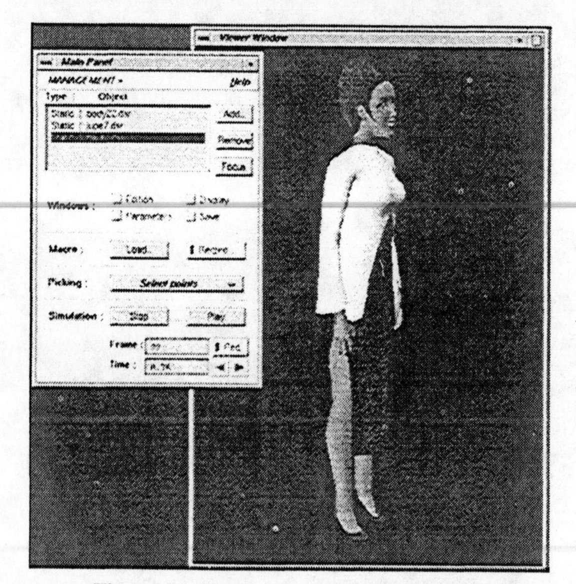

Figure 24. Final garments, after adjustments.

The main improvement from the results of our previous work is the computation time. The results above were obtained on a 150MHz R5000 SGI Indy, and take into account mechanical simulation computation as well as full collision detection and self-collision detection. As soon as the cloth begins to fit the body, collision detection becomes the major computation weight, which reached more than 70% in this example. Implementing incremental collision detection algorithms could reduce this.

Interactive cloth applications represent a wide subject at the crossroads; image synthesis, human animation, cloth design and Virtual Reality. We contribute by providing through this work a fast and robust mechanical simulation system. No more complicated than a simple spring-mass particle system, it however simulates quite accurately an elastic material discretised into any irregular triangle mesh. Associated with Runge-Kutta integration and using numerical error evaluations as damping position and speed corrections, our model is robust, and yet powerful by keeping computation timesteps high. A powerful constraint integration scheme also provides a powerful way of handling collisions, as well as a general support for extra interaction and manipulation tools.

Figure 25. Dressing a virtual actor: from initial panels to animation.

Suited for complicated garment simulations as well as for interactive clothing tools, applications may be extended in any direction of the crossroad. We demonstrate the model's efficiency by some simple examples showing the speed, robustness, flexibility, and adaptability for tracking devices and Virtual Reality applications.

We want to take advantage of this potential to push further in the direction of interactive clothing applications. First, a powerful set of virtual tools would allow us to design, assemble and manipulate garments in a very natural way, enabling us to visually "feel" the fabric material in the 3D space. Then, using our VLNET system [Capin 1997], we are preparing tools and techniques for a collaborative interactive system where distant partners together design and fit a common dress on a virtual being.

4.8. Conclusion

We have described an efficient approach to model the human body for virtual humans/clones based on default adaptive templates body parts. The main advantage of the method is the use of body parts templates that are deformed to create new instances of the same body parts. As the shape deformations are used to automatically update the animation structure of the template to the new instance without additional manipulations, the modeling process is much simpler and efficient than the traditional approach. It provides a powerful framework to create virtual humans or clones ready to be embedded inside VR environments and animated. Some additional advantages of the proposed human body shape modeling approach are:

• versatility: the modeling approaches can be applied and combined according to the given context, depending on the availability of the person to clone and the data we can get from them (photos, pictures scanned from a magazine), also depending on the time available for modeling the shape.
• simplicity: it uses low cost, widely available equipment such as cameras to capture the body parts shape.

Integrating the modeling approach together with efficient and scalable human body animation models, such as the one describe for hands simulation, inside a Networked Collaborative Virtual Environment such as VLNET, dedicated to management of virtual humans evolving and interacting inside a virtual environment, makes possible the development of real-time interactive performances such as CyberDance. The introduction of Deformation LODs (D-LODs) based on a scalable deformation model allows adaptation of the animation time consumption to the requested frame speed. Further adaptation of research should be focused in two directions. The first one regarding body modeling would extend the proposed approaches to any body part, including modeling the hair, and define a versatile framework to extract information from various inputs (pictures, anthropometric parameters, etc.). The second consists of developing strategies in order to dynamically switch between two D-LODs, and control the transition between the two D-LODs.

In order to reach a believable level of realism, simulating virtual bodies is not enough. Accessories such as clothes have also to be integrated in the virtual human modeling and simulation. Traditionally, clothes simulation requires physics based simulation. This usually means slow simulation models, unsuitable for real-time environments . We have shown that the clothes problem is not limited to the simulation of the clothes behavior. We also need to design clothes, manipulate them and dress virtual humans. Although we are not yet able to integrate clothes with virtual humans in real-time environments, our current research on clothes show that we are on the way to be able to interactively design clothes and dress virtual humans. We are also near to having virtual dressed humans evolving in real-time inside a virtual environment.

4.9. Acknowledgements

This research is funded by SPP Swiss research program, The Swiss National Science Foundation and EU ESPRIT Project eRENA (Electronic Arenas for Culture, Performance, Art and Entertainment). The authors thank Christopher Joslin for proof reading this document.

References

Apostoliades, A., 1998, *Character Animation*, http://www.3dcafe.com/tuts/tutcs.htm.

Aubel, A., Çapin, T., Carion, S., Lee, E., Magnenat Thalmann, N., Molet, T., Noser, H., Pandzic, I., Sannier, G., Thalmann, D., 1998, Anyone for Tennis ?, *Presence: Teleoperators and Virtual Environments*.

Boulic, R., et al., 1995, The HUMANOID Environment for Interactive Animation of Multiple Deformable Human Characters, *Proceedings Eurographics '95*, Maastricht, Balckwell Publishers, 337.

Breen, D.E., House, D.H., Wozny, M.J., 1994, Predicting the Drap of Woven Cloth Using Interacting Particles, *Computer Graphics*, Proceedings SIGGRAPH 1994 (Orlando), 365.

Capin, T.K., Pandzic, I.S., Noser, H., Magnenat Thalmann, N., Thalmann, D., 1997, Virtual Human Representation and Communication in VLNET-Networked Virtual Environments, *IEEE Computer Graphics and Applications*, Special Issue on Multimedia Highways.

Carignan, M., Yang, Y., Magnenat-Thalmann, N., Thalmann, D., 1992, Dressing Animated Synthetic Actors with Complex Deformable Clothes, *Computer Graphics*, Proceedings SIGGRAPH 1992, **26**, 99.

Carion, S., Sannier, G., Magnenat Thalmann, N., 1998, Virtual Humans in Cyberdance, *Proceedings Computer Graphics International'98*, IEEE Press.

Carlsson, C., Hagsand, O., 1993, DIVE - a Multi-User Virtual Reality System, *Proceedings of IEEE VRAIS '93* (Seattle).

Chadwick, J.E., Hauman, D., Parent, R.E., 1989, Layered Construction for Deformable Animated Character, *Proceedings Siggraph'89*, ACM Press, 243.

Collier, J.R., et al, 1991, Drape Prediction by Means of Finite-Element Analysis, *Journal of the Textile Institute*, **82**, 96.

Delingette, H., Watanabe, Y., Suenaga, Y., 1993, Simplex Based Animation, *Proceedings Computer Animation'93*, edited by N. Magnenat-Thalmann and D. Thalmann, Springer Verlag, 13.

Eberhardt, B., Weber, A., Strasser, W., 1996, A Fast, Flexible, Particle-System Model for Cloth Draping, Computer Graphics in Textiles and Apparel, *IEEE Computer Graphics and Applications*, 52.

Eischen, J.W., Deng, S., Clapp, T.G., 1996, Finite-Element Modeling and Control of Flexible Fabric Parts, Computer Graphics in Textiles and Apparel, *IEEE Computer Graphics and Applications*, 71.

Ekman, P., Friesen, W.V., 1978, *Manual for the Facial Action Coding System*, Consulting Psychology Press, Inc. (Palo Alto).

Endrody, G., 1998, *Character Modeling and Animation in Alias Power Animator*, http://www.3dcafe.com/tuts/alias1.htm.

Farin, G., 1990, Surface Over Dirichlet Tessellations, *Computer Aided Geometric Design*, North-Holland, **7**, 281.

Forcade, T., 1996, Accessible Character Animation, *Computer Graphics World*, http://www.cgw.com/cgw/Archives/1996/08/08model1.html.

Funkhauser, T.A., Sequin, C.H., 1993, Adaptative Display Algorithm for Interactive Frame Rates During Visualization of Complex Virtual Environments, *Proceedings Siggraph'93*, ACM Press, 247.

Garland, M., Heckbert, P.S., 1997, Surface Simplification Using Quadric Error Metrics, *Proceedings SIGGRAPH'97*, 209.

Haumann, D.R., Parent, R.E., 1988, The Behavioral Test-Bed: Obtaining Complex Bevavior with Simple Rules, *The Visual Computer*, Springer-Verlag, **4**, 332.

Hoppe, H., 1996, Progressive Meshes, *Proceedings SIGGRAPH'96*, 99.

Hsu, W., Hugues, J.F., Kaufman, H., 1992, Direct Manipulation of Free-Form Deformations, Proceedings Siggraph'92, ACM Press, 177.

Hutchinson, D., Preston, M., Hewitt, T., 1996, Adaptative Refinement for Mass-Spring Simulations, *Eurographics Workshop on Animation and Simulation* (Poitiers), 31.

Kalra, P., Mangili, A., Magnenat Thalmann, N., Thalmann, D., 1991, SMILE: A Multilayered Facial Animation System, *Proceedings Conference on Modeling in Computer Graphics*, 189.

Kalra, P., et al., 1992, Simulation of Facial Muscle Actions Based on Rational Free-Form Deformations, *Computer Graphics Forum*, Blackwell Publishers, **2**, 65.

Kalra, P., 1993, *An Interactive Multimodal Facial Animation System*, PhD Thesis nr. 1183, EPFL.

Kang, T.J., Yu, W.R., 1995, Drape Simulation of Wowen Fabric Using the Finite-Element Method, *Journal of the Textile Institute*, **86**, 635.

Kapandji, I.A., 1980, *Physiologie Articulaire*, Tome 1, Maloine SA Editions, 45.

Lafleur, B., Magnenat-Thalmann, N., Thalmann, D., 1991, Cloth Animation with Self-Collision Detection, *IFIP conference on Modeling in Computer Graphics Proceedings*, Springer, 179.

LeBlanc, A., Kalra, P., Magnenat Thalmann, N., Thalmann, D., 1991, Sculpting With the "Ball & Mouse" Metaphor, *Proceedings Graphics Interface '91* (Calgary).

Lee, W., Kalra, P., Magnenat Thalmann, N., 1997, Model Based Face Reconstruction for Animation, *Proceedings MultiMedia Modeling'98*, 323.

Litwinowicz, P., Miller, G., 1994, Efficient Techniques for Interactive Texture Placement, *Proceedings SIGGRAPH'94*, 119.

Macedonia, M.R., Zyda, M.J., Pratt, D.R., Barham, P.T., Zestwitz, 1994, NPSNET: A Network Software Architecture for Large-Scale Virtual Environments, *Presence: Teleoperators and Virtual Environments*, **3**.

Maestri, G., 1996, *Digital Character Animation*, New Riders Publishing.

Magnenat Thalmann, N., Pandzic, I.S., Moussaly, J.C., Thalmann, D., Huang, Z., Shen, J., 1995, The Making of the Xian Terra-cotta Soldiers, *Computer Graphics: Developments in Virtual Environments*, edited by R.A. Earnshaw & J.A. Vince, Academic Press, 281.

Moccozet, L., Magnenat Thalmann, N, Dirichlet, 1997, Free-Form Deformations and their Application to Hand Simulation, *Proceedings Computer Animation'97*, IEEE Computer Society Press, 93.

Moccozet, L., 1996, *Hands Modeling and Animation for Virtual Humans*, Thesis report, MIRALab,University of Geneva.

Mulder, A., 1996, *Hand Gestures for HCI*, Hand Centered Studies of Human Movement Project, Technical Report (http://fas.sfu.ca/cs/people/ResearchStaff/amulder/personal/vmi/HCI-gestures.htm), School of Kinesiology, Simon Fraser University.

Pandzic, I.S., Capin, T.K., Magnenat Thalmann, N., Thalmann, D., 1997a, VLNET: A Body-Centered Networked Virtual Environment, *Presence: Teleoperators and Virtual Environments*, **6**.

Pandzic, I.S., Capin, T.K., Lee, E., Magnenat Thalmann, N., Thalmann, D., 1997b, A flexible architecture for Virtual Humans in Networked Collaborative Virtual Environments, *Proceedings Eurographics 97*.

Paouri, A., Magnenat Thalmann, N., Thalmann, D., 1991, Creating Realistic Three-Dimensional Human Shape Characters for Computer Generated Films, *Proceedings Computer Animation'91.*

Safework, http://www.safework.com.

Sambridge, M., Braun, J., MacQueen, H., 1995, Geophysical Parametrization and Interpolation of Irregular Data using Natural Neighbors, *Geophysical Journal International*, **122**, 837.

Sannier, G., Magnenat Thalmann, N., 1997, A User-Friendly Texture-Fitting Methodology for Virtual Humans, *Proceedings Computer Graphics International '97.*

Schroeder, W., Zarge, J., Lorensen, W., 1992, Decimation of Triangle Meshes, *Proceedings SIGGRAPH'92*, 65.

Sederberg, T.W., Parry, S.R., 1986, Free-Form Deformation of Solid Geometric Models, *Proceedings Siggraph'86*, ACM Press, 151.

Shen, J., Chauvineau, E., Thalmann, D., 1996, Fast Realistic Human Body deformations for Animation and VR Applications, *Proceedings IEEE Computer Graphics International '96.*

Sibson, R., 1980, A Vector Identity for the Dirichlet Tessellation, *Math. Proceedings Cambridge Philos. Soc.*, **87**, 151.

Terzopoulos, D., Platt, J.C., Barr, H., 1987, Elastically Deformable Models, *Computer Graphics,* Proceedings SIGGRAPH 1987, **21**, 205.

Terzopoulos, D., Fleischer, K., 1988, Modeling Inelastic Deformation: Viscoelasticity, Plasticity, Fracture, *Computer Graphics*, Proceedings SIGGRAPH 1988, **22**, 269.

Turk, G., 1992, Re-Tiling Polygonal Surfaces, *Proceedings SIGGRAPH'92*, 55.

Volino, P., Magnenat-Thalmann, N., 1994, Efficient Self-Collision Detection on Smoothly Discretised Surface Animation Using Geometrical Shape Regularity, Computer Graphics Forum (Oslo), *Proceedings Eurographics 1994* (Oslo), **13**, 155.

Volino, P., Courchesne, M., Magnenat-Thalmann, N., 1995, Versatile and Efficient Techniques for Simulating Cloth and Other Deformable Objects, *Computer Graphics*, Proceedings SIGGRAPH proceedings 1995 (Los Angeles), 137.

Volino, P., Magnenat-Thalmann, N., Shen, J., Thalmann, D., 1996, An Evolving System for Simulating Clothes on Virtual Actors, Computer Graphics in Textiles and Apparel, *IEEE Computer Graphics and Applications*, 42.

Volino, P., Magnenat-Thalmann, N., 1997, Developing Simulation Techniques for an Interactive Clothing System, Virtual Systems and Multimedia (Geneva),, *Proceedings VSMM 1997*, 109.

Weil, J., 1986, The Synthesis of Cloth Objects, *Computer Graphics*, Proceedings SIGGRAPH'86, **24**, 243.

Werner, H.M., Magnenat-Thalmann, N., Thalmann, D., 1993, User Interface for Fashion Design, *IFIP Trans. on Graphic Design and Visualisation*, 197.

Wilcox, S., 1997, *Muscle Animation*, http://dlf.future.com.au/scotty/TUTORIAL_2.html.

3D artistic view of Tierra (cf. chapter 2), an environment producing synthetic organisms based on a computer metaphor of organic life in which CPU time is the “energy” resource and memory is the “material” resource. Memory is organized into informational patterns that exploit CPU time for self-replication. Mutation generates new forms, and evolution proceeds by natural selection as different creatures compete for CPU time and memory space. Image courtesy Tom Ray.

The Virtual Great Barrier Reef project (cf. chapter 6) is the first DOME immersive Virtual Reality system combining real-time images, artificial life and role-playing entertainment. Image courtesy Scot Refsland.

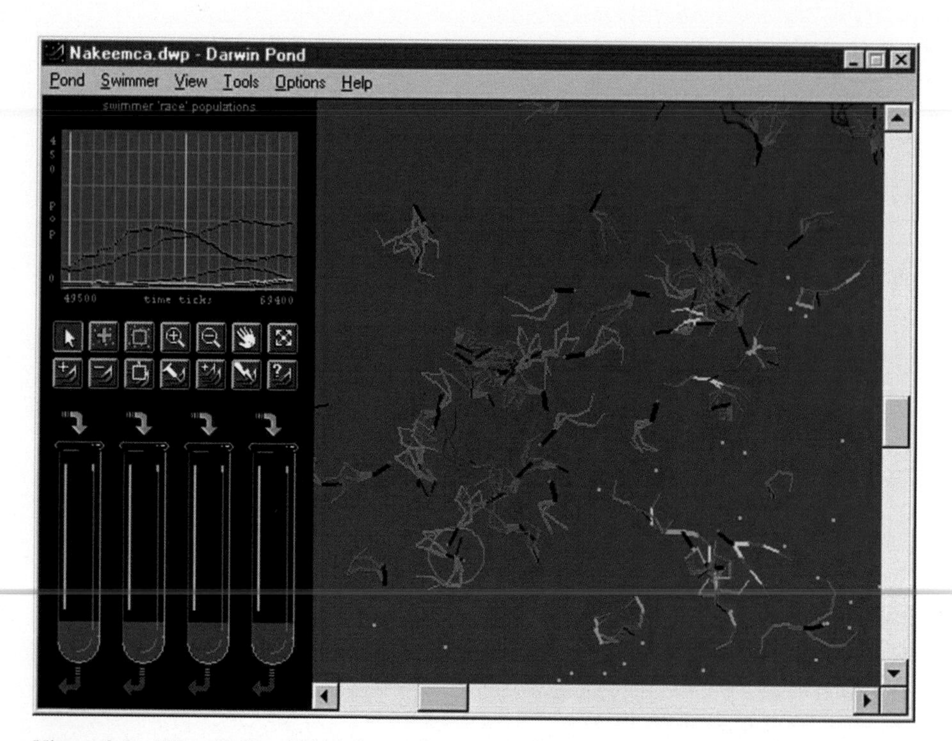

View of the "Darwin Pond" interface (cf. chapter 3). It enables users to observe hundreds of swimmer creatures, composed of 2D line segments and representing a large phenotypic space of anatomies and behaviors. They can be observed in the Pond with the aid of a virtual microscope, allowing panning across, zooming up close to groups of creatures, killing them, moving them around the Pond by dragging them, inquiring about their mental states, feeding them, removing food, and tweaking individual genes while seeing the animated results of genetic engineering in real-time. Image courtesy Jeffrey Ventrella.

A preliminary view of a scene in the Nut virtual world designed at the International Institute of Multimedia (cf. chapter 1). Image courtesy Jean-Claude Heudin.

Plant models generated by the Nerve Garden Germinator including flowering and vine-like forms (cf. chapter 5). These objects are then exported to free "plots" on an island model in a VRML scenegraph. Image courtesy Bruce Damer, Contact Consortium.

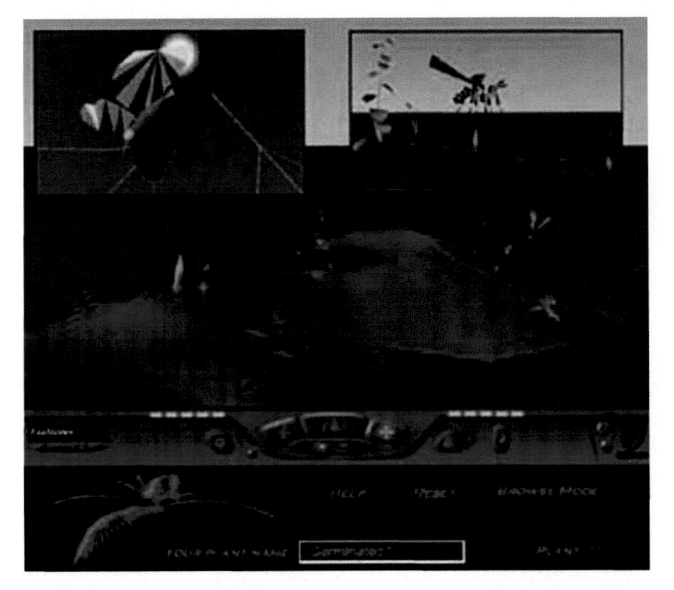

A view of the Nerve Garden interface. Nerve Garden is a biologically-inspired, multi-user, collaborative 3D virtual world available on the Internet: http://www.biota.org/nervegarden. The project combines a number of methods and technologies, includin L-Systems, Java, cellular Automata and VRML. Nerve Garden is a work in progress designed to provide a compelling experience of a virtual terrarium which exhibits properties of growth, decay and cybernetics reminiscent of a simple ecosystem. Image courtesy Bruce Damer, Contact Consortium.

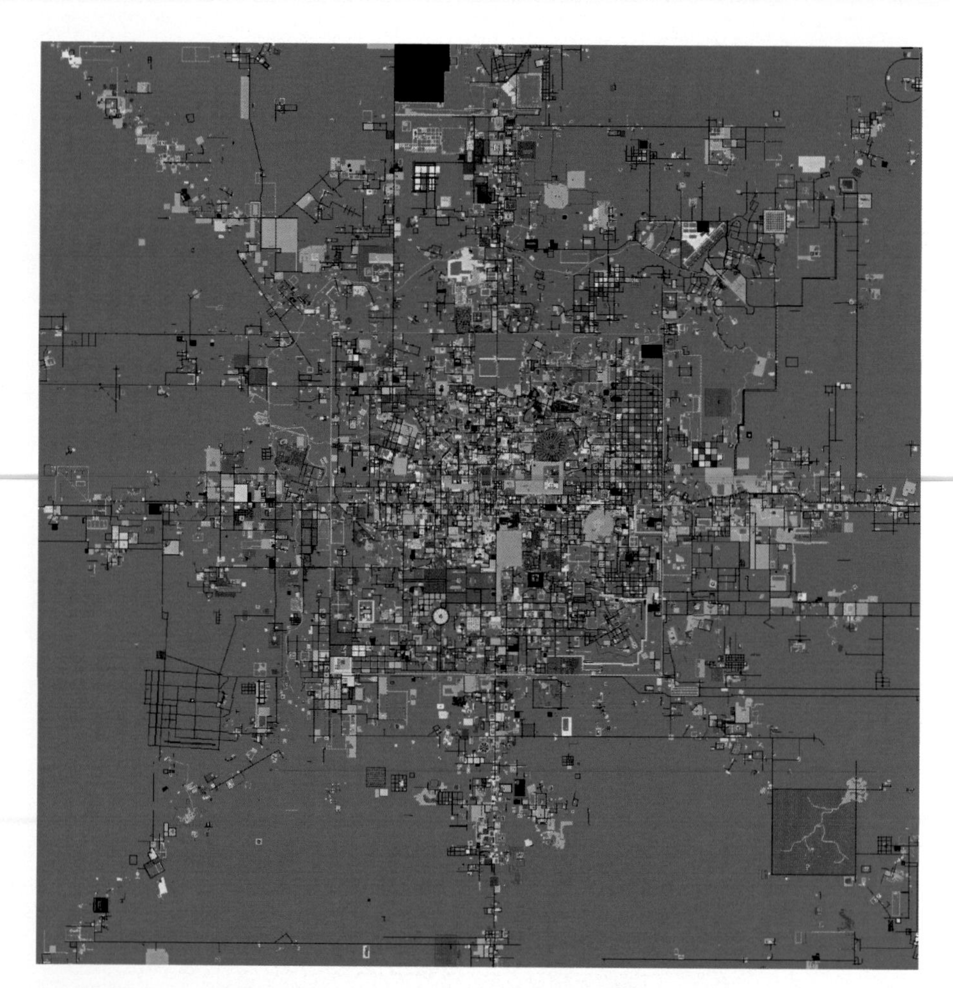

A "satellite view" of the collective construction of AlphaWorld, a huge 3D inhabited virtual city space on the Internet. Some of several hundred thousand visitors to this world placed down approximately 60 million objects since 1995, forming them into the plazas, towns, airports, forests and abstract forms of their visions of cyberspace. Visit this map and AlphaWorld at http://www.digitalspace.com/avatars. Image provided by Bruce Damer, courtesy Roland Vilett, Circle of Fire Studios.

Avatars in the Utopia virtual world available at http://utopia.onlive.com/entrance.sds. Image courtesy Bruce Damer, Contact Consortium.

A view of "Place de la Bastille" in the Second World. An example of a virtual world available to a wide audience on the Internet at http://www.2nd-world.com. Image courtesy Frédéric Le Diberder, Canal+ Virtuel.

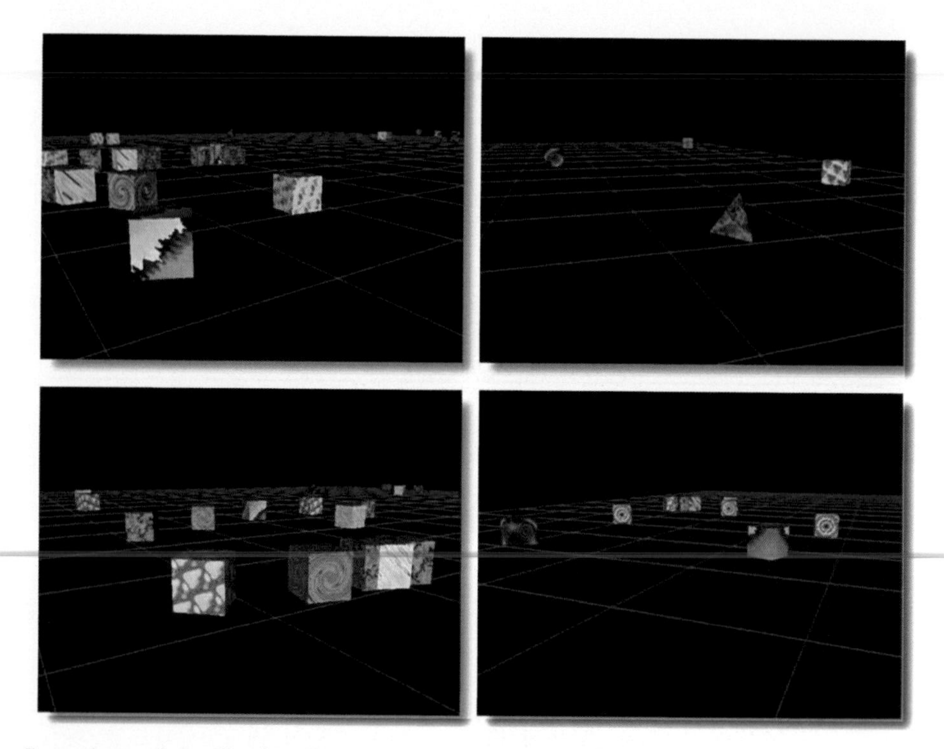

Four views of the Feeping Creatures (cf. chapter 9). This is an interactive virtual world consisting of a population of artificial organisms called feeps. Music and video are continually produced through the feeps' evolution and interactions. It is driven by an aesthetic based on systems and processes rather than objects and images. Nature-inspired processes develop a diversity beyond the direct intentions of the artist. Image courtesy Rodney Berry.

A 3D computer-based artwork designed by David Apikian on the theme of virtual world cities. Image courtesy David Apikian.

"Liquid Mediation" is a narrative virtual environment based on nature, philosophy and architecture. Within the world, an interactant can explore meditative abstractions of natural water reflections. More information about Liquid Mediation is available at http://www.fl.aec.at/~watson. Image courtesy Margaret Watson.

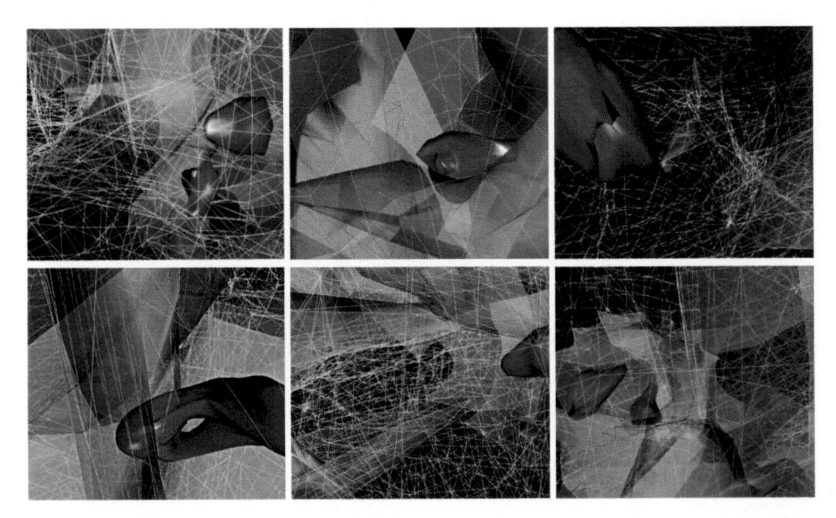

Six views of "Staging Strategies." A virtual world designed using VRML 2.0 by Lilian Jüchterm and Nicole Martin (Programm 5). Staging Strategies is available at http://www.digitalworks.org/rd/p5. Images courtesy Lilian Jüchterm and Nicole Martin, Programm 5.

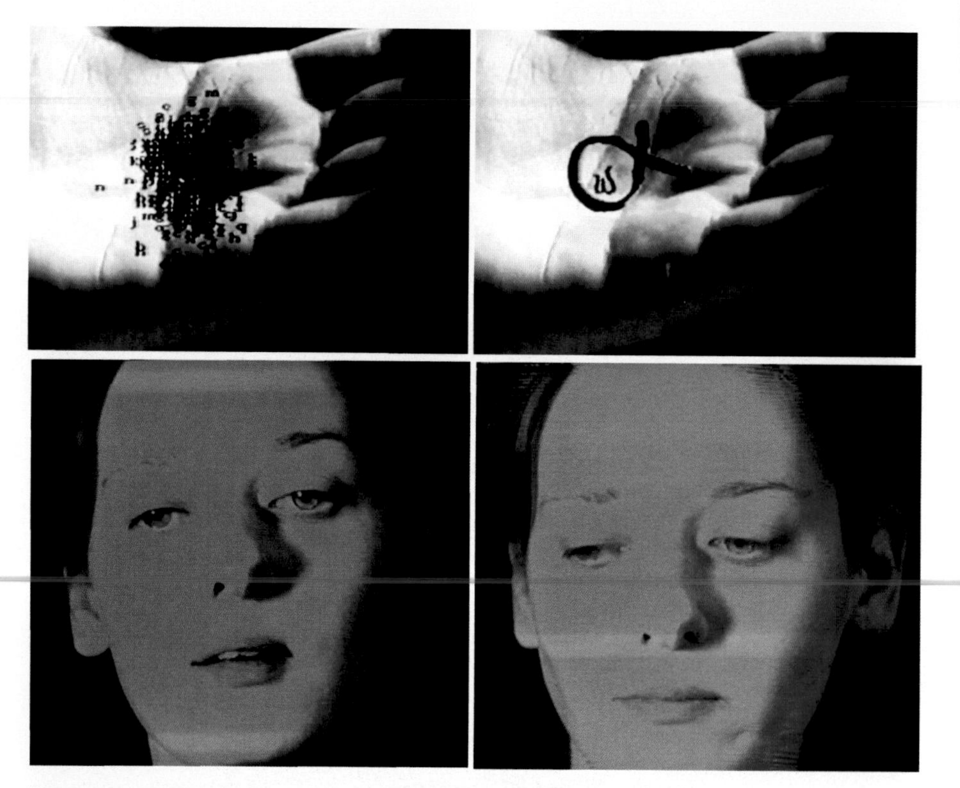

"How are you" (cf. chapter 10) is available at http://www.fraclr.org/howaryou.htm. Image courtesy Olga Kisseleva.

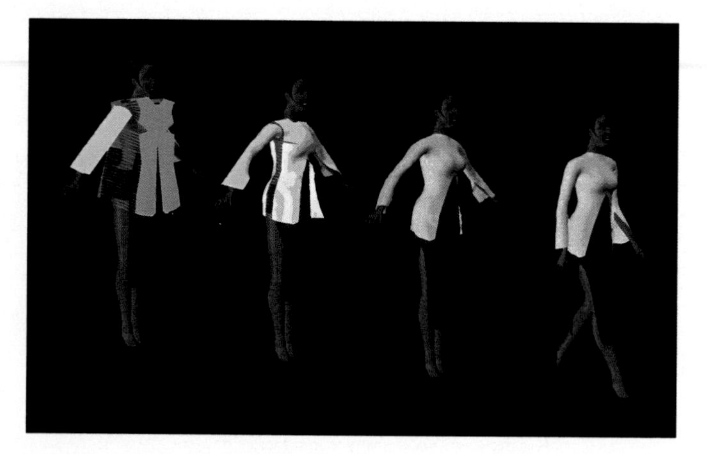

Dressing a virtual actor: from initial panels to animation (cf. chapter 4). Image courtesy Nadia Magnenat-Thalmann, MIRALab, University of Geneva.

Chapter 5
Inhabited Virtual Worlds in Cyberspace

Bruce Damer
Stuart Gold
Karen Marcelo
Frank Revi
Contact Consortium
bdamer@ccon.org
http://www.ccon.org

5.1. Introduction

In his introduction to this volume, Jean-Claude Heudin defines Virtual Worlds (VW) "...as the study of computer programs that implement digital worlds with their own 'physical' and 'biological' laws". He cites two building blocks, VR (concerned with the design of 3D graphical spaces) with AL (focused on the simulation of living systems) and forges a synthesis in VW, which is then "concerned with the simulation of entire worlds and the synthesis of digital universes". This chapter will suggest a third component to the concept of Virtual Worlds, that of the Virtual Communities (VC) that form in and around them. We could refer to this type of environment as an Inhabited Virtual World (IVW).

In this chapter we will briefly review the historical and technical underpinnings of IVSs and then suggest a future role for AL within the online virtual world medium. We will then illustrate some extensive virtual world projects undertaken by the authors' organization, the Contact Consortium, which illustrate principles of emergence in virtual communities and the structures of their worlds. It is useful to point out that the Consortium has its roots in networked learning and contact experiments, so our worlds have a definite pedagogical bent. We will conclude with some short and long term views of where virtual worlds, AL and virtual community may be headed.

5.2. Roots of the Medium

Virtual community finds its technological roots in the earliest text-based multi-user games such as Space War that was a popular application within the early development of Unix. Continuing this trend was the development of UseNET, LISTSERVs, MUDs, MOOs, IRC and conferencing systems like the WELL in the 1970s and 80s [Rheingold 1993], and the World Wide Web and its many progeny in the 90s. The merging of text-based chat channels with a visual interface in which users were represented as 'avatars' occured first in Habitat in the mid 1980s (Benedikt, 1991) and reached an important watermark with the launch of the 3D Internet-based Worlds Chat in the spring of 1995. Strangely, one would suppose that the rise of 'inhabited' 2D and 3D visual spaces in cyberspace would have been heavily influenced by the prior example of Virtual Reality (VR) systems and be closely connected with the development of the Virtual Reality Modeling Language (VRML), but this was not the case. Inhabited Virtual Worlds and their communities drew primarily from their roots in MUDs and text-based real-time chat systems and utilized the power of existing 3D rendering engines developed for gaming applications such as Doom and Quake. VRML has had little to no influence except as an occasional model interchange format and is used in few IVWs with large user communities. IVWs do not take advantage of the full immersion and special devices of VR systems but instead concentrate on running effectively on a large range of consumer computing platforms at modem speeds.

5.3. Technical Underpinnings of Inhabited Virtual Worlds

5.3.1. Technical Underpinnings

Since 1995, a dozen technologies for IVWs have hosted hundreds of thousands of Internet visitors. Each platform represents a distinct vision of what kind of experience a virtual world should offer its communities. This is the 'early adopter' period in the development of the medium and we expect this 'Cambrian Explosion' of approaches to continue for some years. IVWs are still seeking to justify their existence financially, and provide reasons for users to return to the spaces. Some have chosen the game play model. Ultima Online, Quake and other multi player gaming systems have found success by providing a highly structured environment with quests for players to embark on, skills and tokens to acquire, health to endanger, or clans to pit in combat. Social creative virtual worlds, such as WorldsAway or Active Worlds, have relied on inhabitants to build up the community structure, economies, and very content of the spaces. This chapter will focus on these social creative worlds due to their interesting emergent properties and inherently more open architecture. It is this openness that makes these worlds more likely to incorporate metaphors from biology.

5.3.2. How to Build a World?

The technology involved in serving up an inhabited virtual world experience is extensive and impressive. From robust client-server architectures, to streaming 3D object models, to tricks dealing with latency, to citizen authorization and crowd control, and finally to databases managing and mirroring hundreds of millions of objects and thousands of users across networks at modem dial-up speeds, IVWs represent one of the great architectural achievements of computing. We invite you to view the image of one particular cityscape, contained in the color plates section of this book. This 'satellite view' of the Alphaworld cityscape is actually an artful processing of the database of 3D content placed down by some of 200,000 users in the first 18 months of operation from the summer of 1995. Currently over 50 million objects occupy this space, which can be visited by users with ordinary consumer computers running on slow (14.4 or 28.8 KBPS) modem connections.

The literature surrounding virtual world architectures, community development [Damer 1995, 1996, 1998][Powers 1997] and avatar design [Wilcox 1998] is comprehensive and growing, so we will not treat this subject further here.

5.4. Example Virtual World Projects

We will now describe some of the projects in virtual world cyberspace undertaken by the Contact Consortium and its Special Interest Groups (SIGs). These worlds and their resultant communities of interest, were all hosted on line for a general Internet audience between January of 1996 and the fall of 1998. Due to the origins and charter of the organization, these worlds have a distinctly pedagogical bent. The three projects we will review include:

- Sherwood Forest Town: A Virtual Village on the Internet,
- TheU Virtual University Architecture Competition,
- Biota's Nerve Garden: a Public Terrarium in Cyberspace.

Before diving into these worlds, we should consider some of the elements of an effective social matrix needed to make any project a success. Following in the footsteps of successful community and frequent visitation of worlds, a rich set of biological metaphors could be introduced. Artificial Life can both imbue inhabited cyberspace with new meaning and new interest and can also serve as a technology cornerstone in the delivery of the experience to users.

5.4.1. The Platforms Used

At the very birth of this medium, there are a variety of platforms that can support inhabitation over standard Internet connections on consumer computers. There is an exhaustive treatment of these platforms at the "Avatar Teleport" on the web at http://www.digitalspace.com/avatars. We will take a quick review of the major platforms below:

- *Active Worlds*: streaming 3D world for the Windows platform and unix servers with in-world building metaphor. Components are downloaded, cached and replicated. Users build with components with direct manipulation, stacking them like Lego. We employ Active Worlds extensively due to the fact that non-experts, with no knowledge of 3D modeling, can create spaces. Active Worlds can be found at the web at http://www.activeworlds.com.

• *VRML*: a number of projects employ VRML and Java and the External Authoring Interface to create a multi user space. Nerve Garden was one such project. VRML is a powerful general purpose scenegraph language, but requires experts and is not particularly stable or cross platform, running primarily on Windows client systems. The best reference for VRML on the web is the VRML Consortium at http://www.vrml.org. The Blaxxun CCPro product is the most highly developed commercial multi user (avatar) VRML environment and can be found at http://www.blaxxun.com.

• *WorldsAway*: a 2.5 dimensional 3rd person world with a full set of social interfaces, a token object economy, and a well developed community. We use this cross platform (PC and Macintosh) platform for social experiments and group meetings. Find WorldsAway's Dreamscape at http://www.worldsaway.com.

• *Onlive Traveler*: is a 3D voice driven spatialized audio environment for Windows systems which supports lip synchronization with streamed voice over 28.8 modems. We employ this environment for live voice and performance exercises, including musical performances. This world can be obtained from the web site http://www.digitalspace.com/avatars/onlive.html.

• *Other environments*: NTT Interspace, a 3D world for Windows machines that places a video stream onto an avatar and allows voice communications, available at http://www.intsp.com. The Palace, a cross platform environment supporting 2D avatars and simple chat, shared objects, from http://www.thepalace.com.

5.4.2. The Social Matrix

In addition, perhaps more important than the technology of the IVW platforms is the social engineering and people management that strengthens the tenuous web of interactions and shared experience that create communities compelling enough draw users to return to the spaces. Visually impressive 3D effects are not necessarily conducive to good social interaction, as successful 2D environments like Fujitsu's WorldsAway or The Palace have shown. We are only beginning to understand which social user interface affordances promote the development of effective virtual communities. Features such as text or voice chat, private channels, friends lists, muting or expulsion, gift giving, voting and associations, town charters, gestures and many more features are only part of the picture. Best laid plans can do little to encourage or discourage the emergence of social leaders, dispute resolvers, gossip mongers, goodwill spreaders, gatekeepers, community disciplinarians, thieves, abusers and vandals.

Figures of legend, social norms, and virtual world equivalents of urban myths all rise to color the character of virtual communities. A good reference and more rigorous treatment of these themes can be found in the works of Harvard researcher Sherry Turkle [Turkle 1995]. In our three and a half years of anecdotal observations, we have witnessed many fascinating events and heard an array of incredible stories:

Figure 1. A wedding held in AlphaWorld and simultaneously in real life between Tomas (in San Antonio Texas) and Janka (in Tacoma Washington), May 1996.

1. An endless series of weddings between citizens who are getting married for real or treating their cyber betrothals as a whole new form of social interaction (cf. figure 1).

2. Gangs of head banging and boom boxing teens terrorizing Traveler worlds and being recruited into a newly devised team of avatar football players (cf. figure 2).

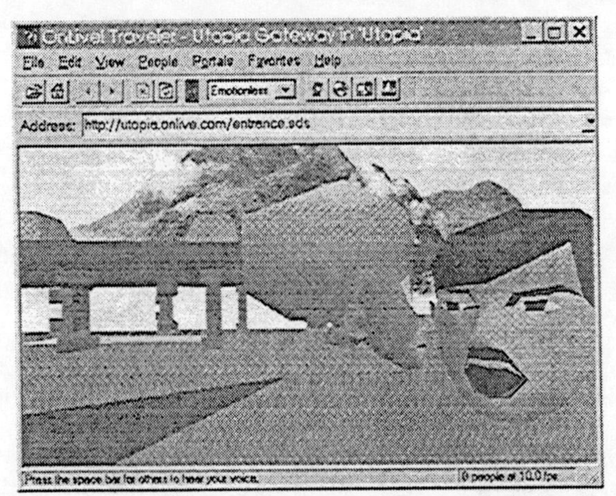

Figure 2. A head-banger attacking a user in Onlive Traveler's Utopia community.

3. An eight month old infant using the microphone to chew and suck his way into contact with avatars in Onlive's Traveler.

4. Rabbis and Zen Buddhist monks resolving a dispute between spiritual center built too close together and learning valuable lessons from each other in the process.

5. The memorialization through virtual gravestones for community citizens, dying either in real life and figuratively (due to evictions from worlds). For example, the death of Princess Diana was marked in cyberspace worlds by flower-bedecked monuments hours before Britons and the rest of the world woke to the news.

Figure 3. An in-person gathering is beamed by webcam into a parallel gathering of avatars, the Reunion, Las Vegas Nevada, July 1998.

6. Other social structures emerge in world including societies, elected officials, and police departments pursuing vandal gangs of former police officers.

7. Face to face meetings of virtual worlds citizens occur and either connect the bonds more securely or shock people with the alternate realities of their real life selves (cf. figure 3).

5.4.3. The Role of Artificial Life in the Inhabited Virtual World

The unique theme of this volume is to investigate how concepts of Artificial Life (AL) are being or could be incorporated into virtual worlds. AL has a unique and powerful role to play in the inhabited virtual world and its communities. The contribution of AL in cyberspace can be divided into two parts:

- To provide biologically inspired behaviors, including animated behaviors, growth and decay, generation and mutation to draw users into these spaces.

- To power underlying architectures with biological metaphors.

5.4.3.1. Using Artificial Life to Draw Attention Span

We have seen the success of non networked CD-ROM games such as "Creatures" from Cyberlife of Cambridge, UK, Petz from P.F. Magic of San Francisco and the ubiquitous Tomogatchi of Japan in capturing the human imagination, attention span and pocket book. Networked environments, such as gameplay systems, and social creative virtual worlds, are all on the verge of acquiring richer biological metaphors. For networked gaming, the drive for more life-like animation, better combatant characters and more rich and changeable worlds inspires efforts such as Motion Factory's Piccolo, a state machine based character animator. Players soon tire of key-framed repeatable behavior sequences and yearn for objects that seem to learn their moves through stimuli from the human players. Believable physics, non-canned motion, stimulus and response learning drive developers to borrow from biology.

Social creative virtual worlds have similar needs to gameplay systems, with less emphasis on real-time low latency action. Moves to introduce biologically inspired experiences are already underway in these spaces. It is felt the introduction of AL metaphors will not only draw in many more users but also strengthen the community matrix. Pets and gardens, perhaps our most intimate biological companions in the physical world, would serve to improve the quality of life in the virtual fold.

5.4.3.2. Artificial Life Powering Better Virtual World Architectures

The recent failure of many efforts in the VRML community to promote an all encompassing standard which would serve behavior rich virtual worlds over the net points out the pressing need for better architectures. The key to delivery of better experiences to a variety of user platforms on low bandwidth connections is to understand that the visual representation of a world and its underlying coding need to be separated. This separation is a fundamental principle of living forms: the abstract coding, the DNA is vastly different than the resulting body. This phenotype/genotype separation also has another powerful property: compression. VRML simply defined a file format, a phenotype, which would be delivered to a variety of different end computers (akin to ecosystems) without any consideration of scaling, or adapting, to that end computer. A biologically inspired virtual world would more effectively package itself in some abstract representation, travel highly compressed along the thin tubes of the Internet, and then generate itself to a complexity appropriate to the compute space in which it finds itself.

As the virtual environment unfolds from its abstraction, it can generate useful controls, or lines of communication, which allow it to talk to the worlds back on servers or to peers on the network. These lines of control can also create new interfaces to the user, providing unique behaviors. One might imagine users plucking fruit from virtual vines only to have those vines grow new runners with fruit in different places. With non-generative, or totally phenotypic models, such interaction would be difficult if not impossible. As we will see in the description of Nerve Garden later in this paper, important scenegraph management techniques such as polygon reduction or level of detail and level of behavior scaling could also be accomplished by the introduction of ecosystem style metaphors. If we define the energy state of a virtual world inversely to the computing resources it is consuming, it would be more beneficial for any scenegraph or objects in it to evolve more efficient representations.

5.5. Sherwood Forest Town: A Virtual Village on the Internet

Sherwood Forest Community (cf. figure 4) was one of the earliest experiment in virtual community building within a 3D world on the Internet. It was carred out between January of 1996 and late 1997 within the first 'constructivist' on-line 3D virtual world: AlphaWorld. AlphaWorld is the first virtual space to be hosted in the Active Worlds environment, now owned by Circle of Fire Studios. Sherwood Forest Town was a district set aside within AlphaWorld of about two hectares in size. The township was built and inhabited by members of the Contact Consortium, an organization dedicated to studying, promoting and enriching Internet-based virtual worlds as a new space for human contact and culture. Consortium members include individuals connecting to the Internet from home, specialists in industry, researchers from universities, and the staffs of companies and government institutions. Consortium members have years of experience in designing and running MUDs virtual communities used in coursework, in computer graphics, and world building exercises. These diverse backgrounds were applied to the Sherwood Forest Community Project.

Figure 4. Meeting in Sherwood Old Town.

The purpose of Sherwood was to design a very natural, attractive setting with woodlands, flowers and flowing water and then attract a community of users to build a village community in that space. A unique feature of Active Worlds is that it allows users all over the Internet with nothing more than a Windows PC and a modem connection to navigate and build in a large virtual space while interacting with others. Using this capability, Sherwood community planners recruited builders from some of the thousands of registered citizens of AlphaWorld. During and after the building of the town, informal observations would be taken and recorded and help the organization carry out more formal exercises later.

Why did we pick the theme of Sherwood Forest? Apart from the attractive fable of Robin Hood (which supplied some imaginative roles), it turns out that the Luddite movement against technology began in the Sherwood Forest region of Britain. We felt that if there was some provocative questioning of (or rebellion against) life in this new virtual worlds technology it might as well happen inside a virtual Sherwood Forest!

5.5.1. Town Charter: Spinning the Social Matrix

Before starting on the building of the site, or the assigning of roles, we created a community charter. Every community needs some sort of charter or constitution or set of rules, whether formal or informal. Sherwood's charter was designed to support the following goal: *To create a viable community of interest within this new medium of human interaction and to observe how this community is built grows and functions.*

The Spirit of our Community Underlying the Charter

- To learn how to work together in a new reality.
- To interact with consideration for others.
- To cherish individual creative independence while meeting the community needs.
- To create a thing of beauty and function worthy of revisiting.
- To have a way cool time.

Basic Community Charter Rules

Be considerate to others and their land and property as you would wish them to be unto you.

Town Services Mandated by the Charter in the Spirit of the Theme of Sherwood

- The town will provide you with newsworthy communication, administration, zoning, and dispute management.
- The town will maintain your mailboxes and deliver your post to others in the community and beyond.
- Townsfolk can give you building instruction and landscaping.
- The town may condemn unused or misused sites.
- The town will clean up trash and provide recycling of objects and reclamation of land.
- The town will publish a free newspaper: the Sherwood Town Crier.
- The town will tow illegally parked cars.

5.5.2. Roles

Some of the roles defined for and filled by participants included:

- *Town founder*: the mythic ancestor DigiGardener who laid out the boundaries of the town and the forests and other features which were later cleared for building.
- *Lady of the Lands*: covered up lots (establishing town domain), placing mailboxes on each, assigning building on lots to others (by uncovering an area and then permitting the invited builder to lay down their own objects).
- *Town Crier*: this 'Robin Hood' character called for town meetings, handled email communications, built the talking circle and other user interfaces to support gatherings, built a website with regular issues (the "Town Crier" news).

• *Greeter*: met people at one of two town gates, arranged and conducted tours, sold visitors on land and building, passing them to the Lady of the Lands.
Building Instructors: helped during "build days" and other times to instruct newcomers on techniques of building.
• *Town Recruiters*: these participants toured over large parts of the adjacent AlphaWorld cityscape seeking to identify quality structures. They would then land and look for a mailbox or other indication or how to contact the builder, reach the builder and seek to bring them into the Sherwood community project. Many highly regarded builders found the small plot size in Sherwood to be challenging. Small co-located plots were sought to enhance intimacy, ease of navigation and prevent the "long walk" syndrome and wayfinding problems of overbuilt virtual worlds.
• *Town Sheriff*: helped arbitrate disputes, controlled security making sure that all lands were covered with tiles (preventing vandalism from the greater AlphaWorld citizenry, who have open access to the site, but not the ability to change objects already placed there). The Sheriff also worked investigated and reported vandalism that did occur in the "New Town" section, an experimental zone outside the town walls which was not covered with protective tiles.
• *Greenthumb*: whose role it was to landscape the properties inside the city walls, to give the town a common feel
• *Village idiot*: stirring the pot, who entered the town when no-one else was there to do trick vandalism (like writing his trickster name 'Kyoti' into unprotected land). This character provided some fun and uncertainty.
• *Psychotherapist*: a real Ericsonian Psychotherapist set up a clinic 'therapies R us' to serve out advice on life in virtual worlds and even provide
• *Storyteller*: this participant documented both the on-line world building and community activities and the physical gatherings held by co-located participants (in Northern California). Tape recordings of interviews of participants, text chat logging and web page construction produced an informal record of the experience and the timeline below.

5.5.3. Sherwood Timeline

From the initial town layout in January 1996 to completion of town construction in July of that year over sixty individuals participated in the project, ranging from 9 year old children to a professional architect and database designer. The following timeline should give you an idea of the phases and events which characterized this experiment:

• *August 1995*: Consortium invited to enter AlphaWorld beta program by developer Worlds Incorporated.
January 1996: Forest and 'ancient aqueduct' boundary defining town wall placed into the world.
• *February 1996*: Web site built, roles defined and filled, community recruitment begins.
• *March 24, 1996*: First big collaborative *build day* happens (cf. figure 5), the talking circle is used to hold an *all avatar* town meeting. A participating Anthro-

pologist makes first attempt to conduct ethnographic interviews of visitors to the town.

• *May 1996*: Second build day fills out the incorporated area of town, Therapies R Us clinic is built, dispute over style of Bazaar built by teenaged citizens arises. Landscape architect "Greenthumb" goes to work on site.

Figure 5. Sherwood citizen co-creator (psychotherapist Steve Lankton) builds Therapies 'R' Us.

• *June 1996*: Prototype virtual university is built in New Town during live exercises by students of Sophia-Antipolis University while at the MediArtech conference in Florence. This event was sponsored by the McLuhan Program of the University of Toronto. Sherwood is avacast between Italy and France and telecast to Italian media.

• *June 8, 1996*: the third build day completes the Old Town incorporated area and plots are completed for the New Town unincorporated areas where freeform building is allowed.

• *July 13, 1996*: Digital singles mixer and summer party gathering is held in Laurel's Herb Farm together with a large physical party in Boulder Creek, California. First ever poetry reading held in Sherwood Redwood Grove. The first eviction of an AlphaWorld citizen was carried out during this event. Read about the event at: http://www.ccon.org/events/mixup1.html. Over 100 avatars participated in the event over an 11 hour period.

• *Late July, 1996*: Vandals struck the unprotected west side properties outside the town walls. A protracted effort to get the vandalized areas cleared began. Proof must be provided to the AlphaWorld city managers that this was not a *creative act*. Call boxes to the AlphaWorld Police Department and Help Patrols are considered for installation at Sherwood for rapid reporting of vandalism and *avabuse* (verbal harrassment of community members by outsiders).

• *August 1996*: More lots inside the New Town area are assigned and building continues. TheU virtual university, a separate virtual world, is created and a teleport transporter system between TheU and Sherwood is put into place.
• *Fall 1996*: A teleport directly to Sherwood is placed at Ground Zero in AlphaWorld. Resultant high traffic encourages vandalism and unplanned seizure of neighbor's land by unknown persons. Areas around Sherwood main gate are filled with teleports to other areas in AlphaWorld. This is likened to the development of low caliber commercial motel zones around Disneyland in the 1950s.
• *March 1997*: Attendees at ACM CHI 97 in Atlanta carry out collaborative building in the New Town area of Sherwood. Sherwood celebrates its one year anniversary. The future of the site and the community is debated and a new land manager sought.
• *July 1997*: Sherwood awarded an honorable mention by Ars Electronica. Published in the Cyber Arts 1997 conference book.
• *August 1997*: Frontage road at Sherwood main gate is hacked and a booby trap teleporter is installed there, firing citizens' avatars far from the site as soon as they arrive. Application made to have the booby trap excised.

Sherwood goes on, and a new round of building and land management is planned. Possible "prop copy" relocation of the Sherwood site to another world is considered.

5.5.4. What Was Learned

The Sherwood Forest Town community experiment was designed as an exercise in collaborative construction, group role playing, goal setting and achievement all in a shared 3D virtual world open to the general public accessible through dial up connections on personal computers. Sherwood was a success in so much as the town was designed, built and maintained for a period of time. The town is still online in fact (see Come Visit Sherwood below). Community roles shifted as the time commitments for Lady of the Land and building instructors were quite extensive. The combination of in-person gatherings with in-world meetings on the major event days was very valuable to keep the teams motivated. Documentation and study of the process, although not rigorous, did provide the framework for a large number of subsequent projects in the Active Worlds environment (we participated in: TheU Virtual University, Arslab, University of California Virtual High School, and the UC Santa Cruz V-Tour amongst others).

5.5.5. Come Visit Sherwood

Find the home of Sherwood Forest Town on the Contact Consortium Homepage at: http://www.ccon.org/events/sherwood.html. Visit the Sherwood Forest Community on the Internet by downloading and installing the Active Worlds Browser http://www.activeworlds.com, entering AlphaWorld, and then teleporting to the coordinates: 105.4N 188.8E (turn around after you land to see the main gate). Note that you can also set up your web browser to teleport directly into various parts of

Sherwood by clicking on teleports found throughout the Sherwood Forest Town Web pages.

5.6. TheU Virtual University Architecture Competition

TheU Virtual University (cf. figure 6) started life as an idea born at a University of Toronto Marshall MCluhan workshop held in the Fortezza da Basso in Florence in May of 1996. A team of people comprising computer professionals and students brainstormed the idea for a virtual university inside a 3D on-line world and created a prototype in AlphaWorld for a presentation at the end of the workshop.

Figure 6. San Francisco State University students meeting in TheU University Development Center.

5.6.1. Birth of an Experimental Pedagogical Virtual World: TheU

Shortly after this workshop, a dedicated Active World server, TheU, was donated to the Contact Consortium for special experimentation. The goal of TheU is to serve as a test bed straddling between traditional campus based universities and the growing number of distance learning projects. Distance learning using current methodology offers many advantages to students in remote areas and students attending part-time courses. However it lacks the sense of community and social interaction which can be achieved by sharing the same environmental spaces and experiences. In the long term Virtual Worlds technology may become a tool for enabling completely new and innovative teaching methods.

We felt we could draw from earlier textual networked learning communities, including SolSys Sim from Northern Arizona University (see it on the web at http://www.nau.edu/anthro/solsys), Diversity University, College Town, VOU (Virtual OnLine UNiversity), to name just a few. TheU would seek to go beyond these experiments due to its unique combination of social interaction, visual human embodiment and user definable virtual environments. Playing it safe for our first experiment, we opted to stay with familiar metaphors of a university campus with its instructional spaces, tutorial help centers, social commons, and library reference zones. Utilizing the power of the Active Worlds builder community, we decided to host a virtual architecture competition to generate a range of approaches to using a virtual world in pedagogy. Fortunately for the organization, Stuart Gold, a British architect and database expert with a long background in online systems, took the lead in this effort, operating the event as a professional juried competition. For your reference, please visit TheU Virtual University and see the results of the Architecture Competition at http://www.ccon.org/theu/index.html.

5.6.2. Competition to Build the First Phase — Institute of Virtual Education

It seemed appropriate that the first faculty of TheU should concern itself with the study and application of Virtual Worlds technology to the field of education. Consequently, the first phase of the university is to be the Institute of Virtual Education. It is hoped that in the early days of applying this new technology, the Institute will attract people from all over the world to debate and take part in the future development of the University and of Virtual Education in general. The Institute's role will be to:

- Stimulate and nurture debate on the development of Virtual Education.
- Provide a research facility and forum for the development of Virtual Worlds technology.
- Provide information and instruction on the use of virtual environments.
- Act as a governing body for TheU, developing and implementing policy decisions.
- Aims and Objectives of the Competition.

Since the inception of Alpha World, its citizens have produced many stunning and innovative virtual spaces and structures. It therefore seemed appropriate to ask Alpha World citizens to take part in a competition, hosted and sponsored by SRT Enterprises and the Contact Consortium, to design the Institute of Virtual Education. The goal of the participants was to create a virtual environment which could best meet the following criteria on which it would be judged:

- Suitability and usefulness as a space to house and facilitate the functions of the Institute.
- Innovative application of Virtual Worlds technology and other emergent technologies such as Web Casting.
- Use of space and objects to create a stimulating, aesthetically pleasing and compelling environment.

Participants were given a large degree of freedom in their interpretation of the requirements. As the driving technology is so fluid and is constantly evolving we felt that the participants should evolve their ideas over the period of the competition and have access to a consultative panel of experts, who would also be the judges, via a listserv available to all the entrants. The panel was made up of educators, architects, artists/designers and technologists including: Murray Turoff from the New Jersey Institute of Technology, Marcos Novak from UCLA, Derrick Woodham from the University of Cincinnati, Gerhard Schmitt from the ETH in Zurich, John Tiffin of Florida State, two of the authors of this paper, and others.

5.6.3. Competition Procedures and Prizes

There were 34 teams, each of whom were given an Active Worlds personal server which was donated by Circle of Fire and hosted by SRT Enterprises. Each of the servers were configured to allow any Active World user to enter and only one participant (or a team using privileged passwords) to build in the world. The duration of the competition was around six weeks.

Due to the open-ended nature of the requirements and the emphasis on ideology and innovation, the participants were encouraged to develop one or more web pages, linked to their schemes, which served to describe their designs and the concepts behind them. Participants were encouraged to outline how their designs related to the learning and educational process and how, if at all, they incorporated or could be extended to take account of, future changes in the technology.

Figure 7. Competition participants, judges and bystanders meet in the pavilion.

A competition pavilion was constructed in TheU (cf. figure 7) which acted as a communal area for all the participants, and contained teleports for Active World users and competition judges to enter each of the worlds.

5.6.4. The Competition

After six weeks of intense building in all 34 competition worlds, six entries were short listed for serious consideration. On March 20, 1998, the final walk through of the finalist worlds with their builders, the judges and a large group (40 to 50) of observers occurred. This event lasted almost four hours. With any in-world event, careful planning must occur to keep it from degenerating into chaos. The following steps were taken to make this coherent, and obtain value for the participants:

1. The event had a clear structure, advertised from the beginning: meeting in the Pavilion, a tour of the six finalist worlds, with the judges, narrated by the builders of the worlds. Strict control of the pacing, moving on between each world every 20 minutes. A final gathering at the pavilion, voting and announcement of the winners by placing banners in the pavilion for all to see.
2. Assignment of gatekeepers and guides: we had trained world users at the pavilion to guide newcomers and folks who crashed their software, back into the live tour. They used private telegram to stay in constant contact with the event leaders.
3. A pair of event leaders, in charge of herding of attendees, crowd control, discipline of event crashers, communications with gatekeepers and generally keeping the conversation interesting and on track.
4. Documentarians were logging all text chat and taking a continuous series of screen shots of the walk through, which can all be seen at http://www.ccon.org/theu/album-background.html.
5. Voting of judges was by secret ballot, sent by telegram and email with followup after the event.

5.6.5. What Was Learned

As TheU competition was seen as a following in the footsteps of the origina Sherwood Town experiment, we felt we had achieved some major goals including: greater consistency of the event, including clear, published goals, well defined roles and concrete outputs: the winning worlds and their documentation.

Did we construct a viable Institute of Virtual Education? Apart from the great aesthetic value of the winning world, Aurac, and its merit for use in demonstrations, it did not serve our needs for the Institute. To take these experiments to the next generation, Bonnie DeVarco lead a team in the construction of a "virtual high school" (VHS) under the umbrella of the University of California at Santa Cruz VHS initiative. This site is hosted in TheU and was initiated with design input by DigitalSpace Corporation (Bruce Damer and Stuart Gold participating). The VHS was built by Craig and Penny Twining of Active Arts Design. Bonnie DeVarco also worked tirelessly to obtain the permission for the display of works of numerous authors and artists made accessible on the walls of the various wings of the VHS. This space contained a great number of images, links, objects and audio narration in virtual tutorial and classroom areas focused on science, language and geometry. You can learn more about the VHS at http://vhs.ucsc.edu/>http://vhs.ucsc.edu. The richness of both TheU competition and VHS environments yielded their own problems: overbuilding created slow frame rate performance and the danger of a "museum effect", in that the environments became static demonstration areas only. Three valuable principles emerging from this experience were:

1. Do not make your virtual worlds too large. Large spaces can cause users to get lost and provide a scarcity of immediate stimulating objects and other affordances to draw them on to the next activity.

2. Do not design worlds that seek to model real world places, unless those places are particularly suited to support interaction in a virtual environment. Navigation and habitation of virtual spaces is so different than the same activities in a physical setting, that a great deal of wasted objects and real estate can result if one is trying to be faithful to the real world setting. Potemkin villages, theme parks, town squares or shopping malls, designed for denser crowding with plenty to see and do are some of the very few real world models worth emulating in virtual space.

3. Design worlds that are constantly changing and changeable. In fact, the "ground zero" (default entry area) of TheU has become a major meeting area, filled with the changing detritus of signage, teleports and weblinks from prior events while the 'museum areas' are static and preserved only for narrative tours.

5.6.6. Come Visit TheU

Find TheU Virtual University on the Contact Consortium Homepage at: http://www.ccon.org/theu/index.html. TheU is a frequently used, changing world that will continue to evolve. Visit TheU on the Internet by downloading and installing the Active Worlds Browser http://www.activeworlds.com, entering AlphaWorld, and then selecting TheU from the listing of worlds on the left hand side of the interface.

5.7. Biota's Nerve Garden: a Public Terrarium in Cyberspace

Nerve Garden is the Consortium's first major attempt to marry Artificial Life metaphors with virtual worlds. The projects described earlier in this chapter were attempts to build a strong social context and achieve something meaningful inside a 3D inhabited space. The Nerve Garden was designed to bring a multi-user biologically inspired space online and eventually marry it with avatar embodied social environments. This is a work in process and we invite your participation.

5.7.1. History of the project

During the summer of 1994, one of us (Damer) paid a visit to the Santa Fe Institute for discussions with Chris Langton and his student team working on the Swarm project. Two fortuitous things were happening during that visit, SFI was installing the first Mosaic Web browsers, and digital movies of Karl Sims' evolving 'block creatures' (Sims, 1994) were being viewed through the Web by amazed students (figure 8 and on the Internet at http://www.biota.org/conf97/ksims.html). It was postulated then that the combination of the emerging backbone of the Internet, a distributed simulation environment like Swarm and the compelling 3D visuals and underlying techniques of Sims' creatures could be combined to produce something very compelling: on-line virtual worlds in which thousands of users could collaboratively experiment with biological paradigms.

One of the Contact Consortium's special interest groups, called Biota.org, was chartered in mid 1996 to develop virtual worlds using techniques from the Artificial Life field. Its first effort is Nerve Garden, which came on-line in August of 1997 at the SIGGRAPH'97 conference. Three hundred visitors to the Nerve Garden installation used L-systems and Java to germinate plants models into a shared VRML world hosted on the Internet. Biota.org is now developing a subsequent version of Nerve Garden, which will embody more biological paradigms, and, we hope, create an environment capable of supporting education, research, and cross-pollination between traditional AL subject areas and other fields.

Figure 8. View of Karl Sims' original evolving block creatures in competition.

5.7.2. Nerve Garden I: Architecture and Experience

Nerve Garden I (cf. figure 9) is a biologically-inspired shared state 3D virtual world available to a wide audience through standard Internet protocols running on all major hardware platforms. Nerve Garden was inspired by the original work on ALife by Chris Langton [Langton 1992], the digital ecosystem called Tierra by Tom Ray [Ray 1994a] and the evolving 3D virtual creatures of Karl Sims [Sims 1994]. Nerve Garden sources its models from the work on L-systems by Aristide Lindenmayer, Przemyslaw Prusinkiewicz and Radomir Mech [Prusinkiewicz 1992][Mech 1996].

The first version of the system, Nerve Garden I, allowed users to operate a Java client, the Germinator (cf. figure 10) to extrude 3D plant models generated from L-systems. The 3D interface in the Java client provided an immediate 3D experience of various L-system plant and arthropod forms. Users employed a slider bar to extrude the models in real-time and a mutator to randomize production rules in the L-systems and generate variants on the plant models. After germinating several plants, the user would select one, name it and submit it into to a common VRML97 scenegraph called the Seeder Garden.

The object passed to the Seeder Garden contained the VRML export from the Germinator, the plant name and other data. Another Java application, called Nerve-Server, received this object and determined a free 'plot' on an island model in a VRML scenegraph. Each island had a set number of plots and showed the user where his or her plant was assigned by a red sphere operated through the VRML external authoring interface (EAI). Cybergardeners would open the Seeder Garden window where they would then move the indicator sphere with their plant attached and place it into the scene.

Various scenegraph viewpoints were available to users, including a moving view-point on the back of an animated model of a flying insect endlessly touring the island. Users would often spot their plant as the bee or butterfly made a close approach over the island. Over 10MB of sound, some of it also generated algorithmically, emanated from different objects on the island added to the immersion of the experience. For added effect, L-system based fractal VRML lightening (with generated thunder) occa-sionally streaked across the sky above the Seeder Garden islands.

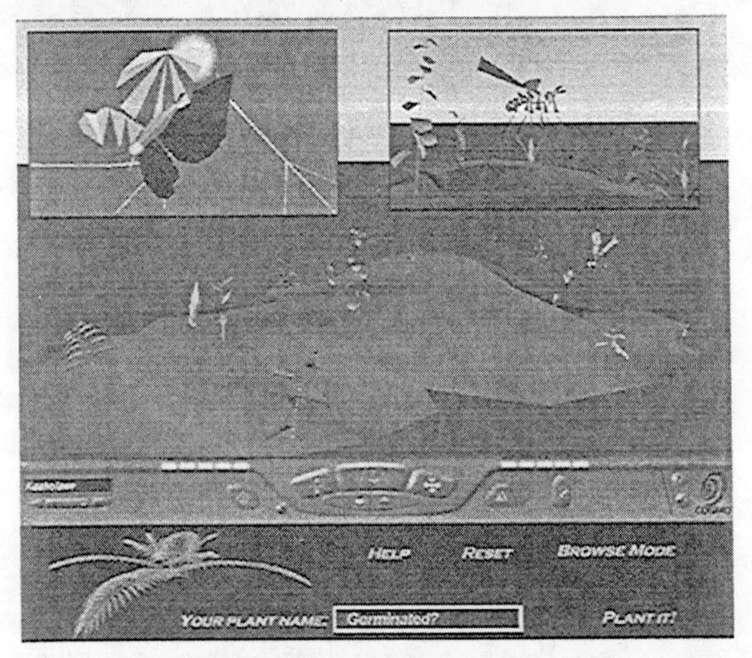

Figure 9. Flight of the bumblebee above Nerve Garden.

NerveServer permitted multiple users to update and view the same island. In addi-tion, users could navigate the same space using standard VRML plug-ins to Web browsers on SGI workstations, PCs or Macintosh computers from various parts of the Internet. One problem was that the distributed L-system clients could easily generate scenes with several hundred thousand polygons, rendering them impossible to visit. We used 3D hardware acceleration, including an SGI Onyx II Infinite Reality system and a PC running a 3D Labs Permedia video acceleration card to permit a more com-plex environment to be experienced by users. In 1999 and beyond, a whole new generation of 3D chip sets on 32 and 64 bit platforms will enable highly complex 3D interactive environments. There is an interesting parallel here to Ray's work on Tierra, where the energy of the system was proportional to the power of the CPU serving the virtual machine inhabited by Tierran organisms. In many Artificial Life systems, it is not important to have a compelling 3D interface. The benefits to providing one for Nerve Garden are that it encouraged participation and experimentation from a wide group of users. The experience of Nerve Garden I is fully documented on the Web at (see references below). Several gardens generated during the SIGGRAPH '97 installation can be visited.

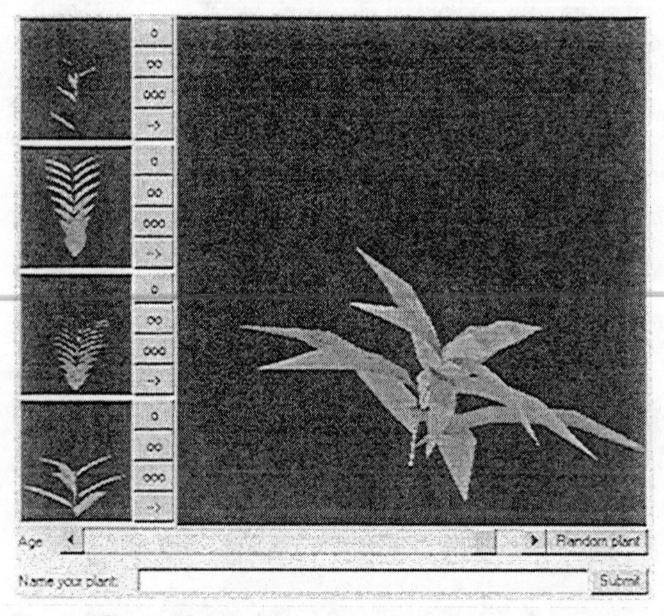

Figure 10. Lace Germinator Java client interface.

5.7.3. What Was Learned

As a complex set of parts including a Java client, simple object distribution system, a multi-user server, a rudimentary database and a shared, persistent VRML scenegraph, Nerve Garden functioned well under the pressures of a diverse range of users on multiple hardware platforms. Users were able to use the Germinator applet without our assistance to generate fairly complex, unique, and aesthetically pleasing models. Users were all familiar with the metaphor of gardens and many were eager to 'visit their plant' again from their home computers. Placing their plants in the VRML Seeder Gardens was more challenging due to the difficulty of navigating in 3D using VRML browsers. Younger users tended to be much more adept at using the 3D environment.

In examination of its deficiencies, while it was a successful user experience of a generative environment, Nerve Garden I lacked the sophistication of a "true AL system" like Tierra [Ray 1994a] in that plant model objects did not reproduce or communicate between virtual machines containing other gardens. In addition, unlike an adaptive L-system space such as the one described in [Mech 1996], the plant models did not interact with their neighbors or the environment. Lastly, there was no concept of autonomous, self replicating objects within the environment. Nerve Garden II, now under development, will address some of these shortcomings, and, we hope, contribute a powerful tool for education and research in the AL community.

In conclusion, did Nerve Garden serve some of the goals for virtual worlds and AL enunciated at the beginning of this chapter? The environment did provide a compelling space to draw attention while also proving that an abstraction of a world, that of an L-system, could be transmitted then generated on the client computer, achieving great compression and efficiency. When combined with streaming and ecosystem controls, Nerve Garden could evolve into a powerful virtual world architecture testbed (see The next steps: Nerve Garden II below).

5.7.4. Visiting Nerve Garden I

Nerve Garden I can be visited using a suitable VRML97 compatible browser. Models made at SIGGRAPH '97 can be viewed at http://www.biota.org/nervegarden. The Biota project and its annual conferences are covered at http://www.biota.org.

5.7.5. The Next Steps: Nerve Garden II

The goals for Nerve Garden II are:

- To develop a simple functioning ecosystem within the VRML scenegraph to control polygon growth and evolve elements of the world through time as partially described in [Mech 1996].
- To integrate with a stronger database to permit garden cloning and inter-garden communication permitting cross pollination between islands.
- To integrate a cellular automata engine which will support autonomous growth and replication of plant models and introduce a class of virtual herbivores ("polyvores") which prey on the plants' polygonal energy stores.
- To stream world geometry through the transmission of generative algorithms (such as the L-systems) rather than geometry, achieving great compression, efficient use of bandwidth and control of polygon explosion and scene evolution on the client side.

Much of the above depends on the availability of a comprehensive scenegraph and behavior control mechanism. In development over the past two years, Nerves is a simple but high performance general purpose cellular automata engine written as both a C++ and Java kernel. Nerves is modeled on the biological processes seen in animal nervous systems, and plant and animal circulatory systems, which all could be reduced to token passing and storage mechanisms. Nerves and its associated language, NerveScript, allows users to define a large number of arbitrary pathways and collection pools supporting flows of arbitrary tokens, token storage, token correlation, and filtering. Nerves borrows many concepts from neural networks and directed graphs used in concert with genetic and generative algorithms as reported by Ray, Sims [Ray 1994b][Sims 1994] and others.

Nerves components will underlie the Seeder Gardens providing functions analogous to a drip irrigation system, defining a finite and therefore regulatory resource from which the plant models must draw for continued growth. In addition, Nerves control paths will be generated as L-system models extrude, providing wiring paths connected to the geometry and proximity sensors in the model. This will permit interaction with the plant models. When pruning of plant geometry occurs or growth stimulus becomes scarce, the transformation of the plant models can be triggered. One step beyond this will be the introduction of autonomous entities into the gardens, which we term 'polyvores', that will seek to convert the 'energy' represented by the polygons in the plant models, into reproductive capacity. Polyvores will provide another source of regulation in this simple ecosystem. Gardens will maintain their interactive capacity, allowing users to enter, germinate plants, introduce polyvores, and prune plants or cull polyvores. Gardens will also run as automatous systems, maintaining polygon complexity within boundaries that allow users to enter the environment.

```
spinalTap.nrv
DEF spinalCordSeg Bundle {
        -spinalTapA-Swim-bodyMotion[4]-Complex;
        -spinalTapB-Swim-bodyMotion[4]-Complex;
        }
```

Figure 11. Sample NerveScript coding language.

We expect to use Nerves to tie much of the above processes together. Like VRML, Nerves is described by a set of public domain APIs and a published language, NerveScript. Figure 6 lists some typical NerveScript statements which describe a two chain neural pathway that might be used as a spinal chord of a simple swimming fish. DEF defines a reusable object spinalCordSeg consisting of input paths spinalTapA and spinalTapB which will only pass the token Swim into a four stage filter called bodyMotion. All generated tokens end up in Complex, another Nerve bundle, defined elsewhere.

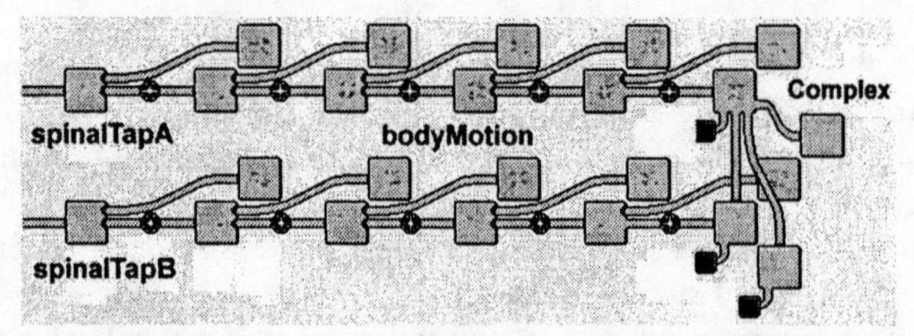

Figure 12. Nerves visualizer running within the NerveScript development environment.

Figure 11 shows the visualization of the running NerveScript code in the NerveScript development environment. In the VRML setting, pathways spinalTapA and B are fed by *eventOut* messages drawn out of the scenegraph while the Nerve bundles generate *eventIns* back to VRML using the EAI. Nerves is fully described at the web address referenced at the end of this paper.

5.7.6. Thanks

In addition to the extensive contributions made by the authors of this paper, we would like to thank the following sponsors: Intervista, Silicon Graphics and Cosmo Software, 3D Labs. A special thanks goes to Przemyslaw Prusinkiewicz and numerous other individuals who have worked on aspects of the project since 1995.

5.8. The Long View: Virtual Worlds as a Primordial Soup for Society, Technology and Life on Earth

5.8.1. The Future of the Medium

Virtual Worlds are a medium in search of an application. At a July 1998 Avatar conference in Banff, Alberta, Canada, a consensus emerged that it was too early in the medium to know how it would ultimately be used. It was felt that it was good that a 'killer app' had not been identified and that avatar cyberspace had time to continue to evolve for its own sake and not to serve possibly inappropriate applications. Comparisons were made to the birth of important technological media of the past century. The telephone was first thought of as a way to distribute music, early film was first cast as a facsimile of theater, and the radio was considered as a method for the delivery of lectures and person to person two way communication between communities.

It is clear from the recent falling off of investment in commercial virtual world platforms, that the medium is entering its "winter", supported only by die-hard users and smaller efforts, largely to build home-brewed virtual worlds and virtual spaces from university and research communities. Millions of computer users have been only recently acclimatized to using classic windows and icons desktop metaphors. Those users are now comfortable in dealing in a cyberspace made up of lists of text, and documents in the form of the Web. Another generation, brought up on Doom, Quake, Nintendo 64 and other environments that stress navigation through very complex, 3D spaces full of behaviors, may be more apt to demand a cyberspace that is built around the metaphor of a place, not just an interface. Will that generation bring us more into virtual worlds for play, learning, work and just being? What will cyberspace look like in ten years, like Gibson's *Matrix* or Stevenson's *Metaverse* [Gibson 1984][Stephenson 1992] or will the document based web and streaming video and audio spaces be the dominant paradigm?

5.8.2. Prospects for a Cambrian Explosion Inside Cyberspace

At the first annual conference of Biota.org, held in the summer of 1997 at the site of the Burgess Shale in Canada, an important source of fossil evidence of the creatures of the "Cambrian explosion" of life 535 million years ago, Tom Ray asked the question: "what are the conditions for the creation of a Cambrian explosion in cyberspace?". Let us finish this chapter with a journey into the farthest realms of speculation. Let us reverse the question and ask "why is life trying to create the conditions for a Cambrian explosion in digital space?". We strive to introduce our concepts of 'artificial' life to a new medium, virtual worlds. If in these spaces we witness the basic metaphors of biological systems, regardless of whether they are expressed as atoms or bits, we may be creating truly new forms of life. Or we may be a surrogate for selfish genes to find their way into a very remarkable new ecosystem with some very special properties. The speculative piece below was derived from a speech given by one of us (Damer) at the second annual Biota conference held at Magdalene College, Cambridge UK in the Summer of 1998. This conference is described at http://www.cyberbiology.org.

5.8.3. Simulacara to Nanocara, or Why is Life Trying to Enter Digital Space?

Kevin Kelly writes in "Out of Control" in reference to the Gaia theory that "...summer thunderstorms may be life raining on itself". He also declares that "...the realm of the born... and the realm of the made... are becoming one. Machines are becoming biological and the biological is becoming engineered" [Kelly 1994].

Seen from another perspective, alien visitors gazing down on our world from high above would almost certainly declare hominid cities, redwood forests, roadways, termite complexes, power grids, coral reefs and other macro structures as all very interesting growths of the biosystem.

Indeed, the grid-patterned, store and forward, clock-pulsed nature of our city streets would be seen by these aliens mirrored in the design of computer circuitry. And yes, every time I fly over Silicon Valley it looks more and more like an Intel Pentium Processor! In addition, at a deeper level, the aliens would note similarities between the molecular token flow machinery of the cell and the message traffic and control linkage of computer operating systems and networks.

There is strong evidence that stromatolites and other ancient cousins transformed the oceans and atmosphere, adding free oxygen and shaping the geology of continents. Nature has once again contrived a powerful biological catalyst agent and is again raining change on itself. But where is Nature headed this time?

5.8.4. Breaching Barriers

If life first developed around deep hydrothermal vents it must have managed to travel through the blackness from extincting smokers to new sources.

The Stomatolites and their ilk opened a much more energetic biotic highway for those cells fortunate enough to metabolize oxygen, the new pollutant of the day. Steven Rooke will talk later about this Oxygen holocaust and Margulis' symbiogenesis. This highway carried life through the water-air-land barriers to our own time.

Barrier breaching required the constant engineering and co-opting of new supporting machinery. Margulis' eukaryotic symbiots absorbed simpler bacteria along the way up the oxygen and solar energy highways.

Then came the hominid brain and its teams of memes driving these ten digits to write yet more supporting structures for the next journey. Cyberspace began as a mesh for memes but where memes tread can genes be far behind? Will the meme, carefully encoded in its new digital bilipid casement, create a new eukaryotic cell by absorbing the more automatous digital biote?

In a brief 20th century lineage of digital biota we have seen the coming of simple computer viruses, digital creatures as pet things, and larger virtual ecosystems like Tierra, hovering around the few rich cyberthermal bitstreams and feeding centers of human attention.

Will we see in the next century a new life cycle: meme gestation, happening for a time in human minds, meme-gene colonization, replication and mutation in cyberspace, light speed travel, and time stasis in dormant storage, and a chrysalis to moth stage spun into atoms and out into the universe by molecular nanofabrication?

5.8.5. But Why is Life Trying to Enter Digital Space?

Can selfish genes in a sort of collective angst be aware that the clock will ultimately run down? Individuals are programmed to die but do not genes strive to continue for ever? Perhaps these genes created consciousness to contemplate life's origins and its ends. Through us, our genes can now know about their ultimate end. The inevitable end will happen in a distant eon as our red giant sun, starved of hydrogen, consumes all genes and memes still dwelling down in this gravity well. Other ends could be more imminent, including untimely impacts or the impact of the prodigious human CO2 emitter. There is evidence is mounting that the Permian extinction (much more deadly than the more famous Jurassic event) was a near miss with a total greenhouse effect. And perhaps not an absolute extinction but a setback of a billion years of machinery and atmosphere building is as good as a termination condition.

So will life enter digital space in search of an ultimate persistence? What advantages are there to be gained by evolving to live in such an ephemeral, narrow and arid ecosystem? One advantage is super charged selection, freed from the slow speed of chemical replication and limited supply lines of atoms, the essential genetic machinery can be copied relentlessly rapidly. Of course, the machinery of the ecosystem itself is still atomic and at this point extended relatively slowly by we the memetic partner. When the atomic support system is also spun by the genes of the digital biotes, the pace will pick up dramatically.

One other distinct advantage of digital biota is lightspeed travel. Massless organisms can escape gravity's well easily, and traverse the solar system in mere hours. If Freeman Dyson is right the collective surface of the Oort cloud, Kuiper belt and various comets and asteroids may be the largest and most fertile surface for ex-terrestrial life.

20th Century emissaries to this zone include spacecraft such as Galileo, whose software brain is constantly morphed and improved by the meme partners at JPL. Clearly humans are poorly adapted to a hard vacuum. Preceding us into these realms will need be countless infrastructure builders better suited for life there.

So will we live to see sometime in the next millenium, biologic nano fabricated asteriodal lichens in symbiotic spore spreading colonies in the outer solar system instanced out of cyberspace?

And at least some of Earth's biological eggs will be out of the proverbial single basket. This would be a wonderful legacy that humankind could leave from its glorious and destructive time in the biosphere.

References

Benedikt, M. (ed.), 1991, *Cyberspace: First Steps,* MIT Press (Cambridge).

Damer, B., Kekenes, C, Hoffman, T., 1995, Inhabited Digital Spaces, *ACM CHI '96 Companion*, 9.

Damer, B., 1996, Inhabited Virtual Worlds, *ACM interactions*, 27.

Damer, B., 1998, *Avatars, Exploring and Building Virtual Worlds on the Internet*, Peachpit Press (Berkeley).

Gibson, W., 1984, *Neuromancer*, Ace Books (New York).

Kelly, K., 1994, *Out of Control: The New Biology of Machines, Social Systems and the Economic World,* Perseus Press.

Langton, C., 1992, Life at the Edge of Chaos, *Artificial Life II*, Addison-Wesley (Redwood City), 41.

Mech, R., Prusinkiewicz, P., 1994, Visual Models of Plants Interacting with Their Environment, Proceedings of ACM SIGGRAPH 96, Computer Graphics, 397.

Powers, M., 1997, *How to Program Virtual Communities, Attract New Web Visitors and Get Them to Stay*, Ziff-Davis Press (New York).

Prusinkiewicz, P., Lindenmayer, A. (eds.), 1990, *The Algorithmic Beauty of Plants,* Springer Verlag (New York).

Ray, T. S., 1994a, Netlife - Creating a jungle on the Internet: Nonlocated online digital territories, incorporations and the matrix. *Knowbotic Research*, **3**.

Ray, T. S., 1994b, Neural Networks, Genetic Algorithms and Artificial Life: Adaptive Computation, *Proceedings of the 1994 ALife, Genetic Algorithm and Neural Networks Seminar*, Institute of Systems, Control and Information Engineers, 1.

Rheingold, H., 1993, *The Virtual Community: Homesteading on the Electronic Frontier*, HarperPerennial (New York).

Sims, K., 1994, Evolving Virtual Creatures, *Computer Graphics (Siggraph '94)*, Annual Conference Proceedings, 43.

Stephenson, N., 1992, *Snow Crash*, Bantam Spectra (New York).

Turkle, S., 1995, *Life on the Screen: Identity in the Age of the Internet*, Simon and Schuster (New York).

Wilcox, S., 1998, *Creating 3D Avatars*, Wiley (New York).

Chapter 6

Virtual Great Barrier Reef

S. Refsland[1]
T. Ojika
Virtual System Laboratory
Gifu University
C. Loeffler
Simation
T. DeFanti
Electronic Visualization Laboratory
University of Illinois at Chicago
Thrane@vsl.gifu-u.ac.jp

6.1. Introduction

The Virtual Great Barrier Reef installation is the first DOME immersive VR system (cf. figure 1) to explore interactivity and complexity on a "many to many" basis. It combines CAVE and DOME technologies together to give visitors a unique experience through a combination of real-time images, artificial life and role-playing entertainment.

6.1.1. Project Overview

In this chapter we discuss the theoretical development of the Virtual Great Barrier Reef installation using proven, existing technologies. The innovation being proposed lies within the integrated complexity of the system and the methodology of the application. This project deals with immense complexity, not only from a machine perspective but also a human factor one. These complexity features are the DOME physical environment, Computer and Machine Systems, real-time connection to the reef through High Definition Vision systems, artificial life and avatars, and interactivity.

---

[1] With also the participation of: C. Lattaud, V. DeLeon, S. Semwal, X.Tu, A. Johnson, J. Leigh,G. Proctor, and R. Gilson.

Figure 1. Virtual Great Barrier Reef Installation.

6.1.1.1. The DOME

The installation is a 15-meter diameter immersive DOME [Virtuarium] Virtual Reality system designed to hold 70-100 visitors interacting with 360 x 360 3D sights and sounds (cf. figure 2). The concept is to create a new style of interactive learning that uses "Nintendo style" role-play entertainment loaded with high levels of factual information and authenticity. The innovations presented are:

6.1.1.2. Computer-less environment

The environment is designed to feel immersive in a story, and make a person feel like they are really in a location, not looking through a computer at a location. It is intuitively interactive through "machine vision" and other input devices that translate participants' movements into computer actions. This way the participants have fewer interactions with the computers and peripherals and more interactions with the story.

6.1.1.3. Real-time Connection to the Reef

An immersive 360 x 270 degree HDTV (High Definition Television) camera system that sits under the water in the reef and transmits back live high resolution pictures in real-time to the DOME. This real image is used as the live background in the DOME and the computer graphic marine life are mapped into it with depth and motion perception. The artificial life graphic fish use pattern recognition to identify the real fish of their own species, then "learn" to emulate behaviors, collision detection, and 3D depth perception. The graphics we have been designing are highly accurate and in many cases it is difficult to tell the real from the virtual life.

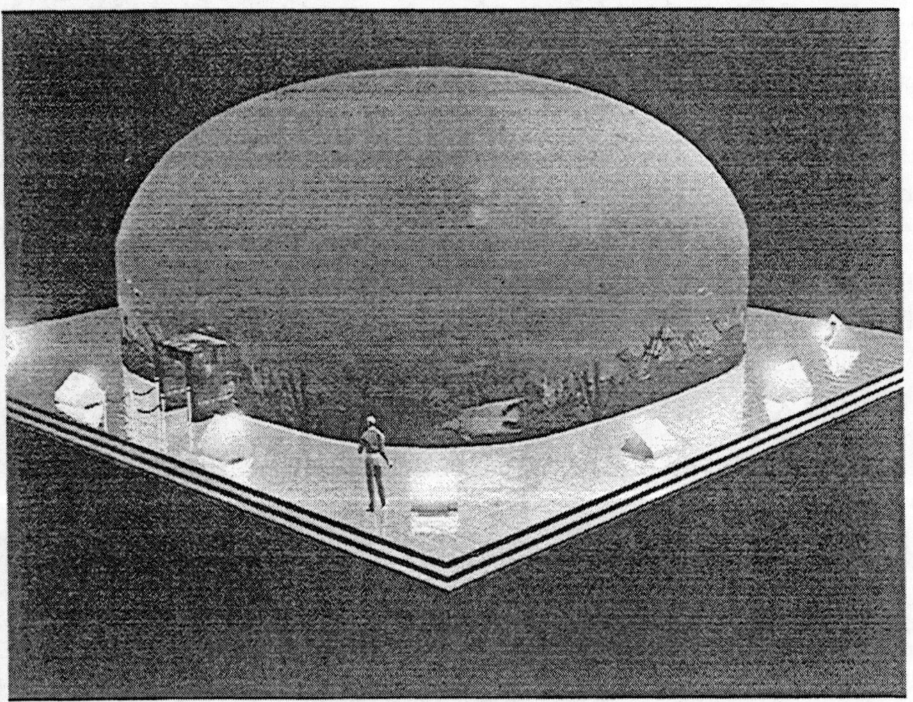

Figure 2. The DOME System.

6.1.1.4. Real Marine Life, Artificial Life, Biobots and Avatars

Through a mix of real marine life being shown through the HDTV system, artifical life fish, biobots and avatars, the entire environment is teaming with all sorts of life, both controllable and uncontrollable. Biobots are avatars that are hybrid – meaning they consist of part human and part artificial life characteristics. This way, participants can role-play various marine life through biobots, and have to behave exactly like the genotype in order to survive. The environment is a mix of artificial life, smaller Biobot avatars and full Motion capture avatars. It is just like a Nintendo game where you start off being a schooling fish and graduate up to larger, more intelligent marine life.

6.1.1.5. Generic Setup

Because of the generic nature of the physical design of the installation, other stories and installations can be shown. We are planning to use the Great Barrier Reef installation as the first of many stories to be interactive DOME capable.

6.1.1.6. Transportable

This entire installation has been designed with portability in mind. It can be transported to almost any facility and built within a reasonable amount of time. Most of the heavy items, such as the computer systems and the projection laser generators are housed within a tractor-trailer and cabled through highspeed networks and fibre optics to the DOME system.

6.1.2. The Components of Complexity

The system is a large distributed network based upon proven, stable DOME and CAVE technologies. Similar reference models in which this project is based upon can be found in the past projects of NICE [Johnson 1998a] and ROBOTIX: MARS MISSION [Robotix].

The NICE project is a collaborative CAVE-based [Cruz-Neira 1993] environment developed at the Electronic Visualization Laboratory at the University of Illinois/Chicago. 6 to 8 year old children, represented as avatars, can collaboratively tend a persistent virtual garden. This highly graphical, immersive virtual space has primarily been designed for use in the CAVE, a multi-person, room-sized virtual reality system. As the CAVE supports multiple simultaneous physical users, a number of children can participate in the learning activities at the same time.

Interactive DOME projects include the Carnegie Science Center's ROBOTIX: MARS MISSION. It was the first example of Interactive Virtual Reality Cinema. Each audience member had a three button mouse device that allowed them to make group decisions on story and piloting spacecraft. The graphics were rendered in real-time, called from an extensive database of Mars's terrain from NASA. The partial dome sat 35 people, and was 200 degrees wide and 60 degrees high. The system used a SGI ONYX, with 4 R10000 processors, Infinite Reality Engine, Audio Serial Option board, and display boards. The polling system, custom built, fed into a PC which talked to the SGI. Majority vote ruled. The dome display used 3 state of the art Electrohome RGB projectors. Most important, only one graphic pipeline was used, and split into three views using custom built software and edge blending. The overlap on edges was 25%, and it appeared seamless and no distortion. Performance, 30-90+ fps, no lag. Audio was 3-D spatialized, custom built software, with floor shakers. It ran for 9 months, no glitches. Effective, there were some people who *thought* they went to Mars.

6.1.3. Methodology of Application

While the true objective of this installation is to install a strong sense of conservation and preservation to the future visitors of the Great Barrier Reef, it seeks to apply it through a Constructionist [Piaget 1973] and "Immersive role-play" educative model steeped heavy in "Nintendo" style interaction of total consciousness submersion. Thus, our methodology objectives are simple – give the visitor an experience that they constructed for themselves, immerse them in wonder and inspiration, in their minds really take them there, and finally, afterwards give them something material they can use to prove and remember their experience.

6.2. Distributed DOME Architectural System

6.2.1. The DOME

The physical DOME is made from an inflatable fabric that is light, durable and highly reflective for adequate projection viewing. Being inflatable has several efficiency benefits, one being light on transportation, ease of setup and dismantling, minimal physical architectural requirements and cooling. Since the DOME is being constantly injected with fresh cool air, it also provides an additional element of altered reality as visitors will go through an airlock and will definitely notice the pressure change. Currently, the projection systems are in 16x9 hi-res format, yielding less projectors per square inch requirements. There are 12 wall projectors, one special 360 degree center floor projector projectong upwards for the inside of the DOME's ceiling, and one 360 ceiling mounted projector to project downward floor coverage.

The DOME machine requirements are constructed of CG and real-time/pre-captured video which is projected onto the scrim of the DOME together in layers to assist with overall processing quality of CG and simulations. The background is a mix of live and pre-recorded scenes in the reef, and the CG fish are compiled into the scene through the projection system to create the mixed illusion.

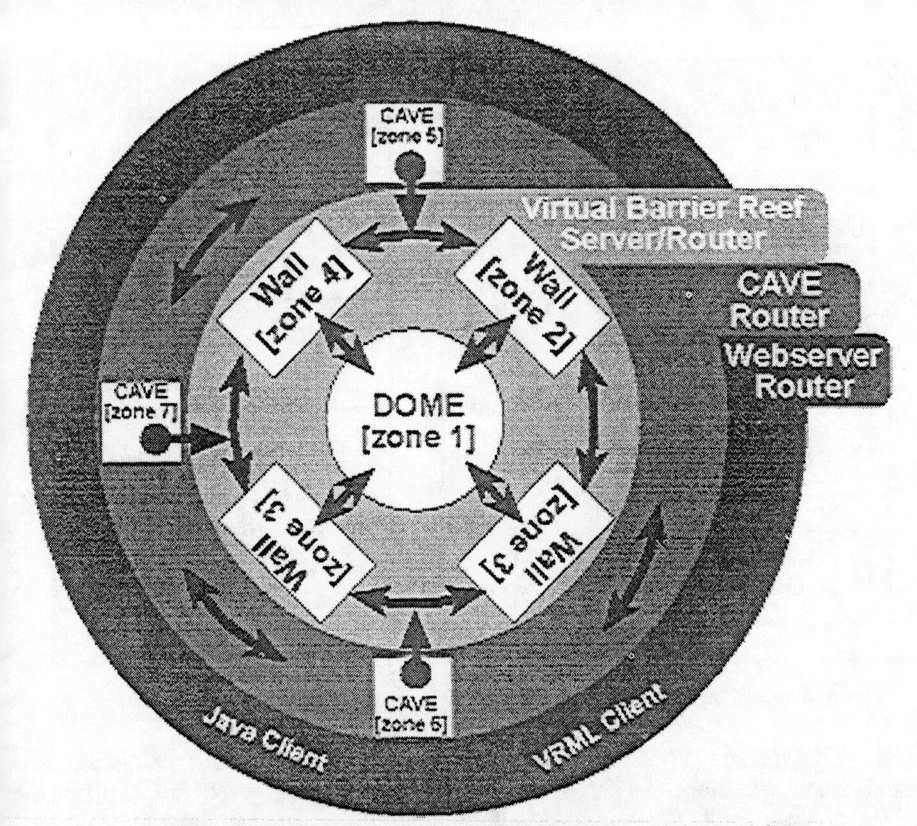

Figure 3. The WALL and DOME Topology Connectivity model.

6.2.2. The WALL Zones

The walls will be the most computational expensive of all the zones in the environment as most of the direct interactivity will happen around them. Walls are interactive in that they might have species that will react when visitors walk by them. Example would be an alife moray eel might come out to feed, and when a visitor walks up to it quickly, a sensing device gives it spatial reaction information. Spot projectors and smaller independent computer systems are displaying various small interactive displays, while the meta environment was separate and being processed by the machine system.

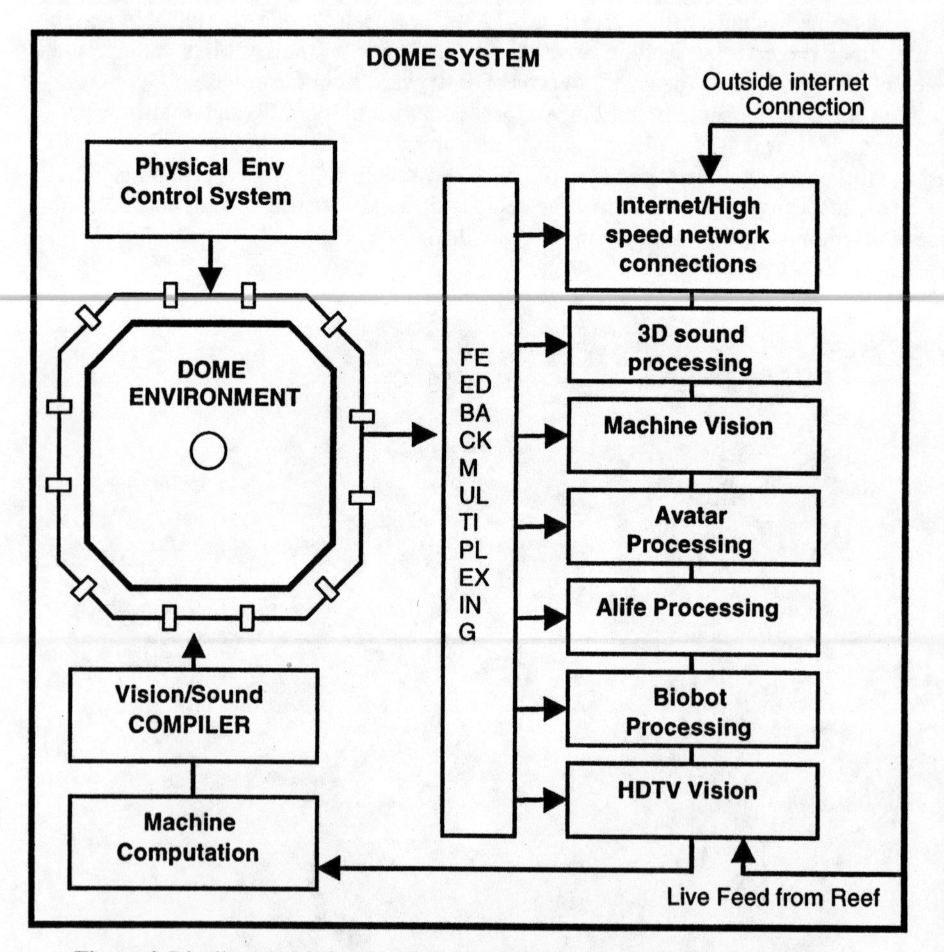

Figure 4. Distributed Relationship of the DOME, WALLs, CAVES, and the Internet.

The DOME and WALLs use the standard CAVE network topologies and protocols being developed as part of the CAVERN network [Cruz-Niera 1993][Leigh 1997] to deal with the multiple information flows maintaining the virtual environments (cf. figure 3).

Each of the WALLs, and DOME zones making up the environment will be connected via a high-speed network, allowing information such as avatar data, virtual world state / meta data and persistent evolutional data to pass between them. This will allow Alife, biobots and avatars to move to any of the WALLS or the DOME, and allow museum guides in the guise of reef inhabitants to interact with the visitors. This will also allow remotely located CAVE users in the United States, Europe, and Asia to join into the collaboration. Similar connectivity has been used in the NICE environment [Johnson 1998b] to link multiple CAVE sites in the United States to Europe and Japan.

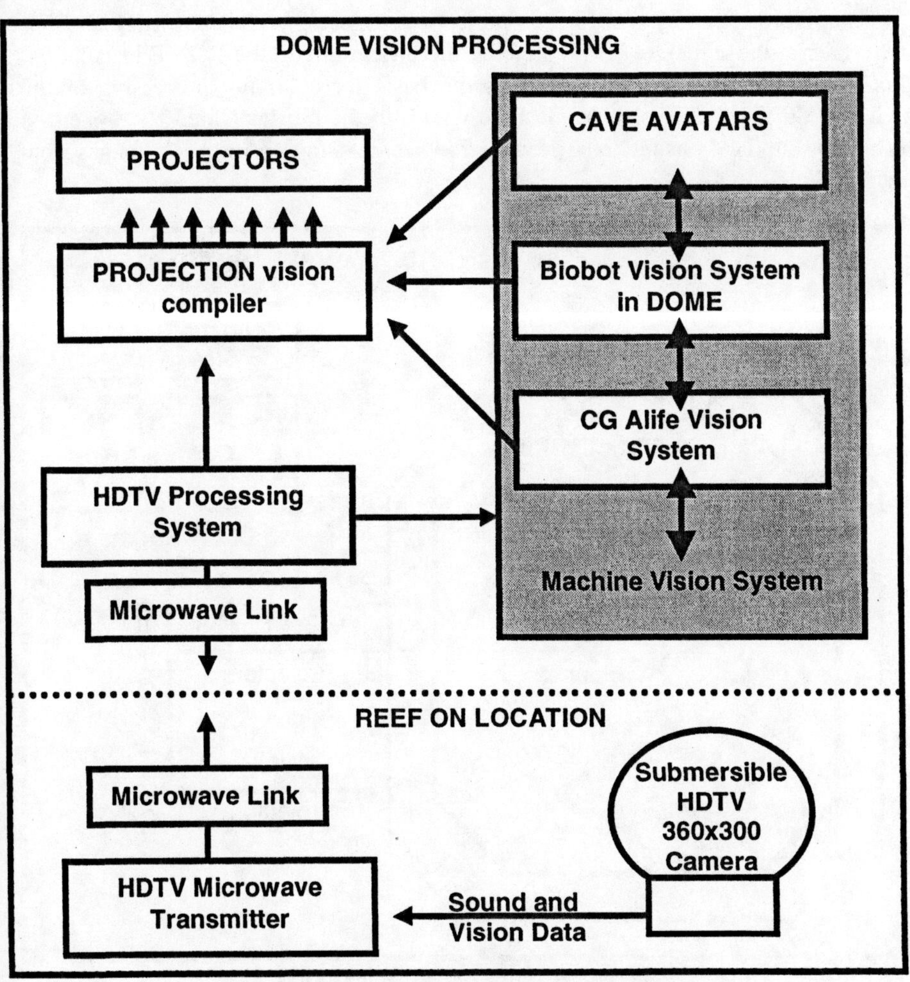

Figure 5. Direct HDTV connection to the real reef.

6.3. Mixed Reality Environment

The environment of this project uses real-time video, lighting, sound and machine vision techniques to establish an environment that convinces the user that they are indeed immersed in the Great Barrier Reef (cf. figure 4).

6.3.1. Real-time Connectivity to the Reef

A major innovation to this project is the mix of virtual reality and real life to create a complex environment. Research is currently being undertaken into a High Definition TV camera/projection system for this project (cf. figure 5). The camera system sits underwater at the actual reef, with a direct microwave link to the DOME installation. This real-time connection to the reef is the basis for all major processing of the environment, from an evolvable machine world, to the artificial life. The system is fairly robust and easy using high speed microwave links to transmit back images from the reef.

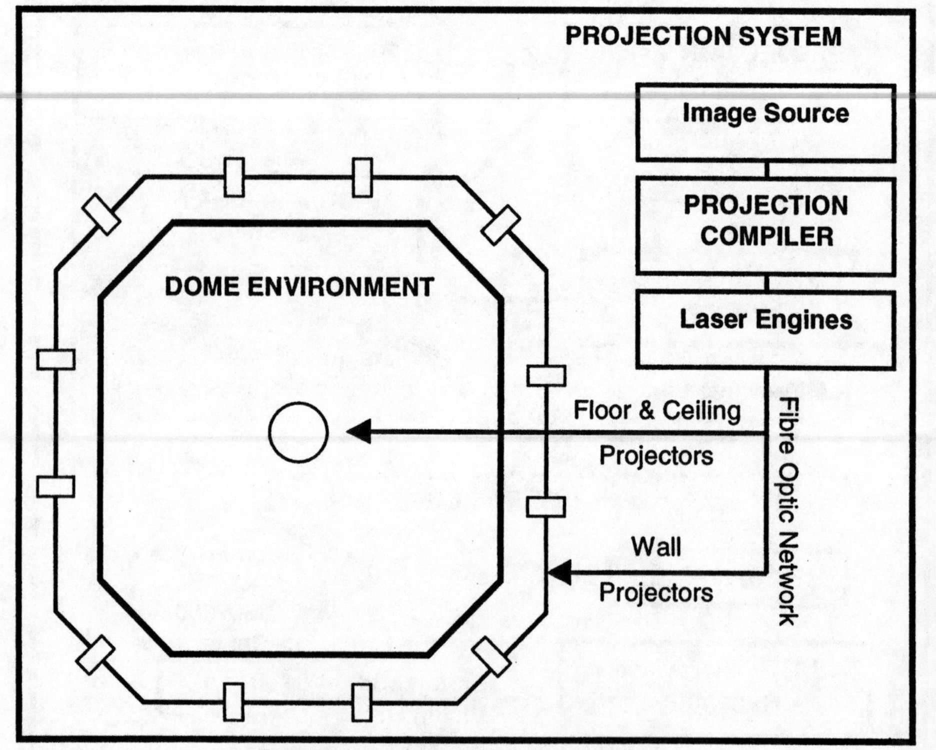

Figure 6. Projection system for the DOME.

6.3.2. Laser Projection Systems

Another innovation this project is researching is a new advancement in projection systems. This system uses laser projection systems to deliver the HDTV feed and CG

data. Since there are many projectors, the need for highly accurate calibration is required (cf. figure 6).
The laser projection systems are ideal for this project as the lens gate system and laser light sources have been separated, making the projection housing extremely light. The laser light sources are in fact in a tractor-trailer outside, being coupled through fibre optics with the lens gate housings around the DOME installation.

6.3.3. Sensing, Tracking and Machine Vision

This section discusses new uses of tracking virtual environments. In fact, it uses a substantial amount of machine vision in order to compile the components of Real marine life, AL, Biobots and avatars together in an interactive fashion. Once these components have been threaded together, they begin a relationship that makes the environment evolvable and effectable from any perspective.

6.3.3.1. Monkey See, Monkey Do and Learn

This project explores not only the relationship of chaotic data in which to give the AL it's temperament, but it uses machine vision in which to provide "real-life" teachers for the AL to learn from. We are researching the notion of using machine vision to track the actual position and behavior of real marine life. When compiled through machine vision, the AL are matched with a HOST, or a real marine life fish that matches its genotype. Once this comparison is done and a match is made, the AL fish then follows and emulates this fish as if it was a child, or "monkey-see-Monkey-do", mimicking and imitating the behaviors of the real fish. By using this HOST information, the AL begins to learn real patterns, then over a period of generations, the AL fish has a fairly accurate behavior pattern of its genotype.

6.3.3.2. 3D Underwater Tracking of Marine Life

In this Section, we explain methodology that can be applied for mimicking marine-life surrounding a VR-Dome in VGBR project. This is a new area for 3D tracking as, for the first time, marine-life as well as participants must be tracked in a virtual environment. In this section, we discuss why camera-based tracking is most promising mode of 3D tracking for marine life. We believe that underwater 3D-tracking would be an essential component for new and exciting artificial-life experiences. At the University of Colorado, Colorado Springs, we are developing the Scan&Track Virtual Environment which is ideally suited for such an application as it provides camera-based tracking using a promising new technique called the active-space indexing method. We also present salient features of this system in this Section.

6.3.3.3. Machine Vision Background

Virtual environments pose severe restrictions on tracking algorithms due to the foremost requirement of real-time interaction. There have been several attempts to track participants in a virtual environment. These can be broadly divided into two main categories: encumbering and non-encumbering [Sutherland 1968][Rashid 1980][Brooks 1990][Krueger 1991][Meyer 1992][Gottschalk 1993][Sturman

1994][Semwal 1996][State 1996]. A variety of encumbering tracking devices have been developed, for example acoustic [Meyer 1992], optical [Rashid 1980] [Gottschalk 1993], mechanical [Sutherland 1968][Brooks 1990][Semwal 1996], bio-controlled, and magnetic trackers [State 1996][Capin 1997].

As indicated by [State 1996], mostly magnetic trackers have been used in virtual environments as they are robust, relatively inexpensive, and have a reasonable working volume or active-space. However, magnetic trackers can exhibit tracking errors of up to 10 cm's, if not calibrated properly [State 1996]. Magnetic trackers are also affected by magnetic objects present in the surroundings. Optical trackers are encumbering as either four dedicated cameras are worn by the user on top of their HMD for the inside-out system, or a set of receiving beacons are placed on the user's head for the outside-in systems [Meyer 1992][Geiger 1996][Semwal 1998].

6.3.3.4. Camera-based 3D Tracking

Most of the camera-image based systems are examples of un-encumbering technology [Krueger 1991][Cruz-Neira 1993][Geiger 1996][State 1996][Capin 1997][Wren 1996][Semwal 1998]. The well-known correspondence problem is ever present when purely vision-based systems are used. As mentioned in [Cruz-Neira 1993], there are several compromises made to resolve the issue of correspondence and recognition. Even then, it is difficult to find a topological correspondence between the multiple contours extracted from the multiple camera-images of the same scene. In the field of computer vision and photogrammetry, there is a wealth of research on motion tracking [Serra 1988][Grimson 1990][Maybank 1993][Faugeras 1996]. There has been much research effort in the area of understanding and recognition of facial expression, body movements, gestures and gaze of the participants [Cruz-Neira 1993]. However, correspondence and occlusion still remain grand challenges.

The resolution and accuracy of camera based systems is dependent upon the size of the pixel. It is difficult to differentiate between any two points if the size of the pixel is large and the projection of both points falls on the same pixel. In our Scan&Track system, we use multiple points in 3D-space to minimize this problem.

Camera-based techniques are well suited for developing un-encumbering applications as the tracking information extracted from the images can be transported far-far away and surround the

VGBR participants and thus can mimic the world around them. This area is well studied, and Krueger's work [Krueger 1991][Geiger 1996][Capin 1997] is well known for virtual environment interaction using 2D contours or silhouette of human participants, and interpreting them in a variety of ways. When only 2D camera-images are used, there is no "depth" information available with the image. Much research has been done in the area of computer vision [Serra 1988][Grimson 1990][Maybank 1993][Faugeras 1996], stereo-vision [Faugeras 1996], object-recognition, and motion analysis [Sutherland 1968][Brooks 1990] to recover the depth information. The active-space indexing method maintains the 3D-Space information, and thus provides a new paradigm for camera-based 3D tracking.

Tracking methods should also be scalable by allowing the size of the working-volume to change as needed. In addition, varying degrees of accuracy and resolution should be available. For example, tracking methods should allow higher accuracy and/or resolution for a high-end (faster) system, yet allow the same 3D-space to be tracked with lower accuracy and resolution for a low-end (slower) system. It is possible to arrange the cameras in the Scan&Track system as desired, and thus, the working volume can be decreased or increased as needed.

6.3.3.5. Why Use Camera-based Tracking in Underwater Applications?

Underwater tracking poses further difficulty for 3D tracking. Both the magnetic trackers and ultrasound trackers may experience robustness, accuracy and reliability problems due to water-contact and possibility of leaks which may occur as the trackers are "directly" placed on the marine-life being tracked. Optical tracking may create reliability and accuracy problems due to refraction and movement of the water and the marine life. Camera-based tracking has distinct advantages in this regard, as cameras can be placed at convenient spots, for example inside a 3D-dome [Cruz-Neira 1993]. And they are more portable and easier to set-up. The workspace or coverage for the camera-based systems is large, in comparison to optical based tracking, and can also be conveniently varied.

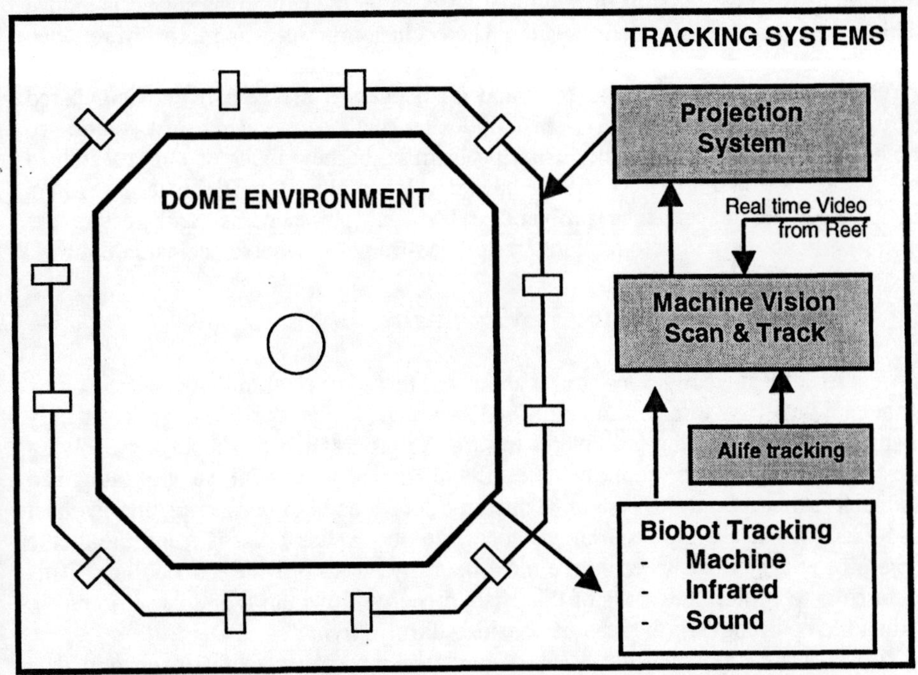

Figure 7. Machine Vision Diagram for tracking real marine life and participants.

6.3.3.6. Why Scan&Track System for Underwater 3D Tracking?

The GVBR project requires that the marine-life is tracked and represented as avatars and/or autonomous independent synthetic quasi-representation of the marine-life. For the VGBR project, one of the experiments we would like to perform is to try our new 3D-tracking algorithm for underwater tracking using HDTV images (cf. figure 7).

We are developing the Scan&Track Virtual Environment [Semwal 1998] at the University of Colorado, Colorado Springs, in cooperation with ATR Multi-media integration and Research Laboratory, Kyoto, Japan[2]. The system uses multiple cameras to track multiple participants and is still under development. The Scan&Track system uses a new method for 3D-position estimation based upon the active-space indexing method. An implementation of the active-space indexing method has been developed. In addition, the geometric-imprints algorithm extracts significant points from camera-images. Active-space indexing algorithm calculates the exact location of a 3D point given the 3D-point's projections on the multiple images. The grid-data structure, used by the active-space indexing method, isolates and provides minimal set of points that correspond to each other. This set is two to three order of magnitudes smaller, providing a fast method for solving correspondence across multiple images.

The occlusion problem is inherent in video-based tracking. How can I know that the participant's hands are in his her pocket? We must track previous frames and see where possibly the hand can go. There is no video-based tracking system that resolves occlusion successfully yet. Since 2D-camera images are used to determine the participant's position, it is difficult to infer whether the participant's hand is hidden behind their body, or is in their pocket. The occlusion problem is more severe when multiple participants are present in a camera-based virtual environment.

Occlusion problems are indeed grand challenges in the area of unencumbered virtual environment. Since active-space-indexing method provides a fast solution to the correspondence problem, occlusion problem is the new focus of our research in the Scan&Track system. Recently we have developed algorithms to create fast filters to solve the correspondence problem across multiple cameras, and results are promising. Now, we plan to look into past video-frames to resolve occlusion issues.

6.4. Complex and Evolving Environment

In weather, climatic conditions are "chaotic" but there is a rhythm that occurs (Lorenz attractor, butterfly effect [Lorenz 1963]) which keeps the weather constantly changing, never repeating yet always toward the progression of supporting evolving life. These conditions profoundly affect life and is beyond control, yet still integrated /interactive in an eco-cycle system. This same sort of naturally occurring energy must also exist within a networked environment, i.e. the web/internet is a large climatic system of energy flow, where there must be a rhythm to provide it's stability. This research is to explore the areas of this chaotic energy flow, tap it and use it to create virtual life/environments that can be considered truly "living".

This research explores the areas of emergent properties found in chaotic data streams occurring within a networked environment. It uses this live chaotic data streams in which to establish emergent rhythmic behaviors to control autonomous life and evolvable environments. As an example, this process gives the artificial life/environments its "pulse". This "pulse" then provides internal/external influences for motivation within the environment, migration to other worlds, and interactions between other life/worlds. The whole environment should be interconnected so that

[2] More details also available at http://www.cs.uccs.edu/~semwal, under NSF98.

all actions affect the whole of the environment as in the real world. Areas of exploration include: enabling alife characters to dream, be consciously aware, provide emotions, have "inspiration", etc. Evolvable environments to have their own eco systems. Both Alife and environments directly integrated for evolvable feedback.

6.4.1. What is a Live Chaotic Data Stream?

If you took the stockmarket, and used the data that is generated everyday, and used something like a Lorenz attractor, you'd find a rhythm, but never a pattern. Another example if you measured the distance between a whale's footprints, or where it surfaced, you would see a rhythm, but not a pattern. Weather conditions are rhythms, but not patterns. Net usage around the world has a rhythm from how the data is being transferred around the net. The human heart beats in an irregular rhythm - and in most cases when the heart falls into a pattern, the person dies (pattern equals death). These are all examples of live chaotic data streams which it is my thought that this is where we are going to find emergent life as it can provide a "pulse" to the organism, and give it a rhythm but not a pattern.

6.4.2. The Environment: A Universe of Many Worlds, Times and Spaces

This research is exploring the direct nature of evolvable worlds and evolvable artificial life. For the Virtual Great Barrier Reef Project, each Wall of the DOME is considered a different world and each of these worlds interact with each other on a meta scale.

In this evolvable environment, each time a new Alife species is born it is allocated to live in a particular area and given high levels of territorial traits. In each world there are many types of Alife all living in the same space, but not necessarily in the same time perspective. For instance some fish live a very short time span where others have a very long cycle. The organism is defined by typical AI genetics, but to provide it a constant variable stream of changing data, the organism is directly affected by the actions of its host. The HOST is the real-time image of a recognized pattern of a real marine species that matches the genotype of the artificial life species genotype. The system begins building a profile of the Host that the Alife recognizes and uses this constantly changing data to affect the organism.

Once organisms begin to populate the world, then the feedback from interactions begin to change, sculpt and shape the world. For instance, Participants behaving like “humans” in the environment will cause green clouds to appear in the water, too many participants in one area at one time will make the area murky and difficult to see. Each of the worlds also have a theme and have particular laws. These laws are slowly changing over time, and are controlled by chaotic data streams from the net.

From a participant level, they will see their actions are directly affecting the bio-mass as a whole. They will also see over a period of time that their organism is starting to look similar to other organisms in color, speed, flocking, time.

6.4.3. Immigration and Valence Individuals

In some cases, some of the organisms can develop singular or "valence" personalities where they are strong/intelligent enough to migrate to other worlds without having to flock. The machine vision will watch for new patterns to emerge within the real-time landscape and eventually if it recognizes a moving pattern that is not in it's database, then it will attempt to build a new artificial creature closely simulating the pattern, movement and behavior of this new unidentified species.

6.4.4. Environmental Triggers

Climatic Interference: The virtual model is connected real-time to a section of the real reef so that it can automatically adjust the virtual model. Temperature, current, air pressure and other climatic conditions on the reef will cause several actions in the virtual environment, such as temperature of the room, flow of the virtual debris, virtual water currents, etc. This environment is considered to be a "living entity", possessing an authentic as possible simulation to the real reef.

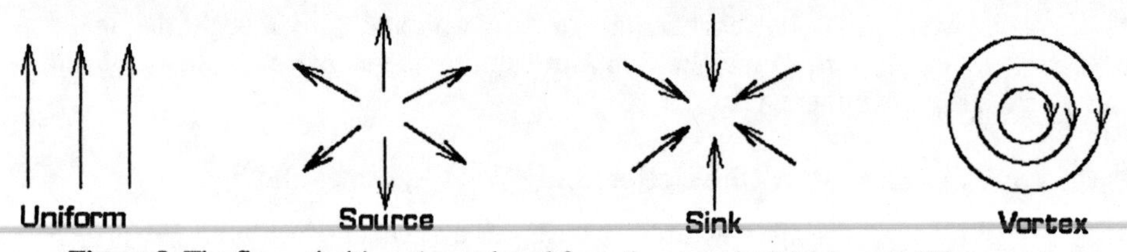

Figure 8. The flow primitives (reproduced from the paper by Wejchert and Haumann).

For simulating simple fluid flow, we consider techniques similar to the ones used by Wejchert and Haumann Wejchert 91 [Wejchert 1991] for animating aerodynamics (cf. figure 8).

Assuming inviscid and irrotational fluid, we can construct a model of a non-turbulent flow field with low computational cost using a set of *flow primitives.* These include *uniform flow* where the fluid velocity follow straight lines; *source flow*–a point from which fluid moves out from all directions; *sink flow* which is opposite to the source flow; and *vortex flow* where fluid moves around in concentric circles. This process is also directly linked with the real reef current flow that is constantly supplying real-time information. Both of the processes integrated together enable lower computation requirements and increase the authenticity of the virtual environment.

Much of the seaweed, plankton, coral and other movable plant/marine life are simulated and respond dynamically to water currents and movements of other life in a realistic manner. In this case it is proposed to use a system which builds the plants in a mass-spring chain assembly. In order to obtain computational efficiency, we do not calculate the forces acting on the geometric surface of each plant leaf, rather, we approximate the hydrodynamic force at each mass point.

6.5. Real Life, Biobots, Artificial Life and Avatars

Generally, most Artificial Life (AL) research is focussing upon evolvable properties within a pre-defined set of rules and environments, where the parameters are preset and the environments are run to feedback upon themselves within a closed environment. This process leaves nothing to chance, or for events to happen outside the researcher's control. It may define the AL's DNA structure, genotype, or behaviors, but it does not provide the "pulse" of life that real life forms get everyday from the natural environment, food, outside events, interactions, observance, variance, and happenings that are "by chance". It also does not consider "bio rhythms" which all natural life is governed by; i.e. moon, tidal, seasonal, species, and other natural cyclic rhythms.

Our treatment of artificial life in this project is an exciting and innovative one. As mentioned earlier, one of the educative goals in this project is to enable people to "think like a fish" so they can experience first hand through the "eyes of a fish" what it's like to live on the reef.

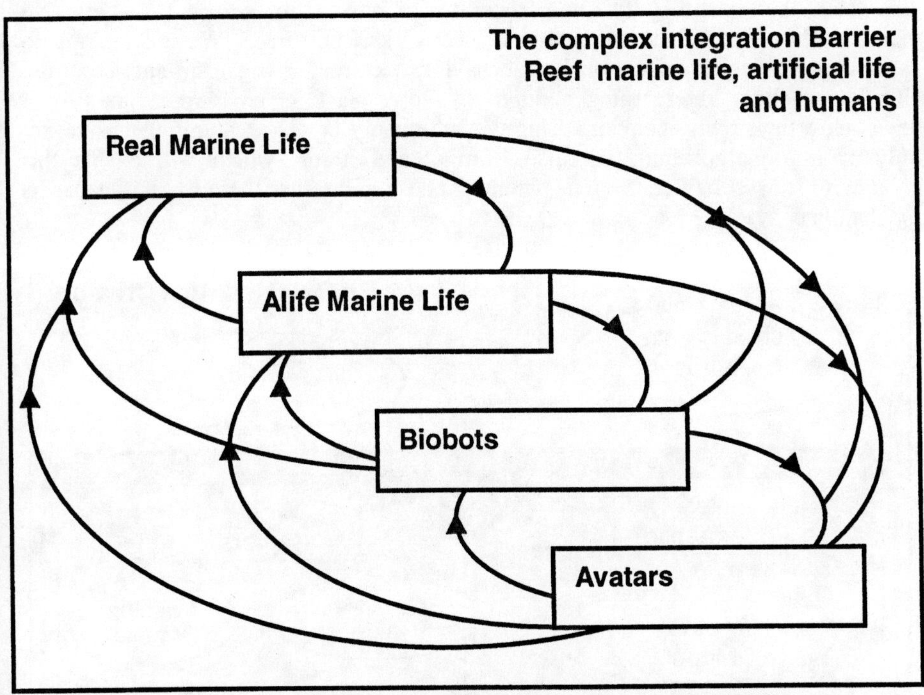

Figure 9. Complex relationship from Reallife to a life back to real life.

To achieve this, we have developed in integrated system for the complete cycle of real-life to artificial life (cf. figure 9). As you can see from figure 9, there is a direct and indirect relationship between the real marine life and the participants. Through the use of artificial life, Biobots and avatars, participants can learn the authentic behaviors of the real marine life as if the participants were actually the marine life themselves. The fundamental model closely modeled to the real marine life is the Alife marine life, or a BioRythmed.

6.5.1. BioRythmeds: Artificial Life With a Chaotic Pulse for a Heartbeat

The AL are a totally autonomous animal that behaves according to the events in the environment. They have a life span, are integrated into the ecological system, and possess authentic behaviors of its real life counterpart. They are closely related to existing Alife models, except that there are two new conditions added, 1) Integrated to other types of artificial life, and 2) directly tied to a real-life genotype counterpart.

6.5.2. BioRythmed Sex and Mutations

All AL will have typical sex and mutation methods built into their behavior systems. What is unique about this system is that we are proposing that when an artificial life creature has offspring, for an incubation period it will use its parent's chaotic information in order to quickly establish its identity and behavior patterns. The organisms have an incubation period, where they use their parent's chaotic stream to give them their pulse, with DNA encoding to have them search for alternative methods, they find other taps within the networked universe. We are looking to explore new methods of establishing behavior: a computer organism emulating the HOST organism, but learning enough to evolve past its existing behavior into developing their own behavior. This will ultimately create creatures that have not only evolution algorithms, but also learning algorithms, which will enable this creature to learn behaviors from its parent and host, and re-use them when it leaves its incubational system.

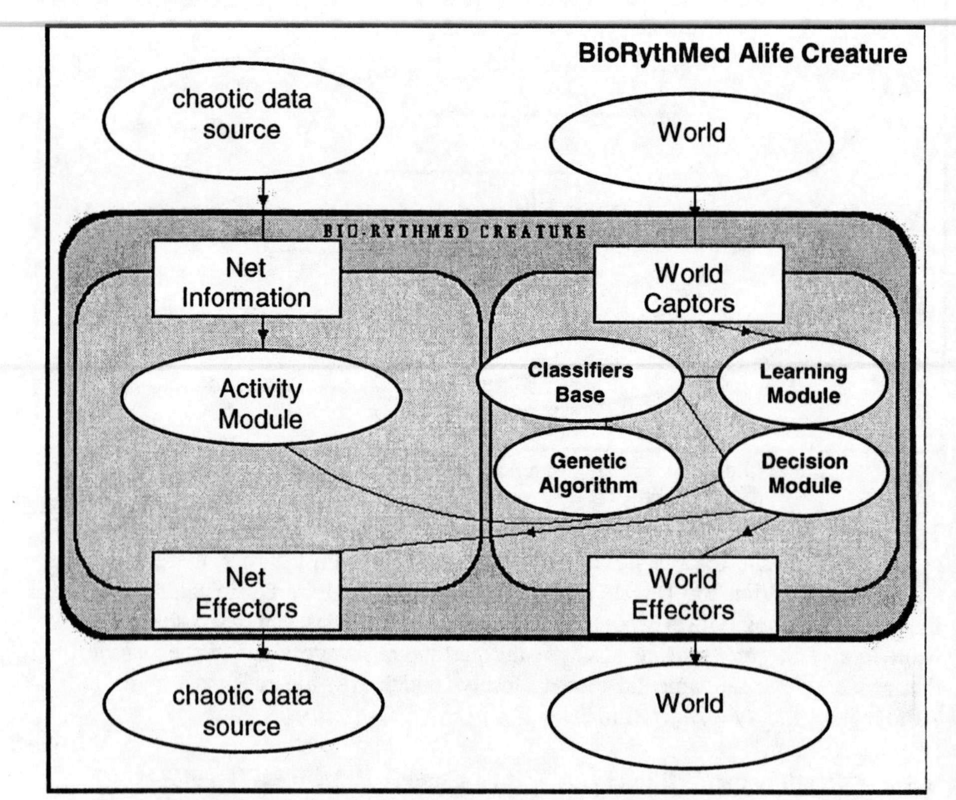

Figure 10. Design of a BioRythmed Creature that is connected to a live datasource.

The behavior of a BRC is defined by a hybrid kind of a Classifier System [Holland 1992], connected to a particular Activity Module. Figure 10 shows the internal structure of a BioRythmed Creature. The activity module checks the activity of its host or the Real marine life that the machine vision has identified for the AL to emulate and learn from, and produce changes in the behavior of the creature. Activity modes can be: explore (reinforce many moves), feed (reinforce feed actions), mate (reinforce mate actions), regroup (reinforce when the creature is near other creatures of the same type), hibernation (reinforce inactivity)... Each of these mode reinforce some sets of classifiers according to their aims. For example, when the host leaves the simulation, the creature can swap to hibernation mode; if the host has a huge activity, the creature can enter into an exploratory mode. Then, for each mode, the creature develops an adapted behavior.

In the BRC model, 3 kinds of captors exist: Environment captors bring information about the world (perception, environmental state), internal captors define the mental state (memory, planning capacities...) and Net captors for internet information. The first two are used to define the classifiers base of the creature and so its behavior, the third defines in which mode the creature is in (and so its behavior too, but at a higher level). Creatures can be conceived with a componential model, where it's easy to add or remove components of agents (like captors and effectors). The hybrid CS is based on the Macro-Mutation operator described in [Lattaud 1998]. This operator, added to the standard operators of the Genetic Algorithm [Goldberg 1989], allows BRC to mute not only genes of their classifiers but also the structure of these classifier. In the reproduction process, a creature can undergo this kind of mutation, and alterate the architecture of its genotype, with respect of its parent's ones. In a phenotypic point of view, this mutation can be interpreted as a deep modification of the creature: It can gain, or lose, captors or effectors, as new legs, wings, eyes and soon. Practically, this operator brings a deeper adaptation of creatures to their environment, and according to initial survival conditions, some emerging behaviors, individual -as obstacle avoidance and following- and collective -as ecological niches-, emerge.

6.5.3. Biobots: Artificial Animals for Life-like Interactive Virtual Reef Creatures

The second model is a similar model, only that the "brain and motor skills" are detached and placed into the control of a visitor using a dataglove to direct the fish, making it a Biobot. The innovation here is that even though a visitor is controlling the "will and motor" of the artificial fish, it still has behavior and characteristic traits that will not allow you to act outside of the character set for that fish.

An example: you are a schooling fish and (you identify your fish amongst a 1000 other fish in the school by a clicking device that you can 'flash' your fish) its behavior laws say it must follow the school in a swarm lest it be eaten if it is alone. If you think as a person, you might go on your own to look at the clam. But as soon as you are away from the school, another autonomous/avatar fish comes up and eats you. Game over! So to survive, you must learn quickly to be a schooling fish.

What's the key to bringing you a captivating experience as a tourist of the virtual Great Barrier Reef? Whether you are merely observing the schools of sardines, playful dolphins and vicious sharks, or you are living the aquatic life by playing the role of one of the virtual reef dwellers (want to be a dolphin, parrot fish or maybe a reef shark? Sure you can), realism is the password to the ultimate immersive experience. Here we are not just talking about visual authenticity in the appearance of the digital sea creatures, but also in their physical environment, and most importantly, in the way they move, perceive and behave. In an interactive environment like ours, behavioral realism, especially, autonomy of the virtual creatures is indispensable.

How are we to achieve such integrated realism? Our answer is to build Artificial Life Biobots (or AL) models of the aquatic animals that are controlled to a degree by human motivation, but also have behavior patterns built into the genotype to override actions that are not found in the charactistic of the marine life. This way, the visitors controlling a Biobot will quickly learn the marine life's realistic behaviors.

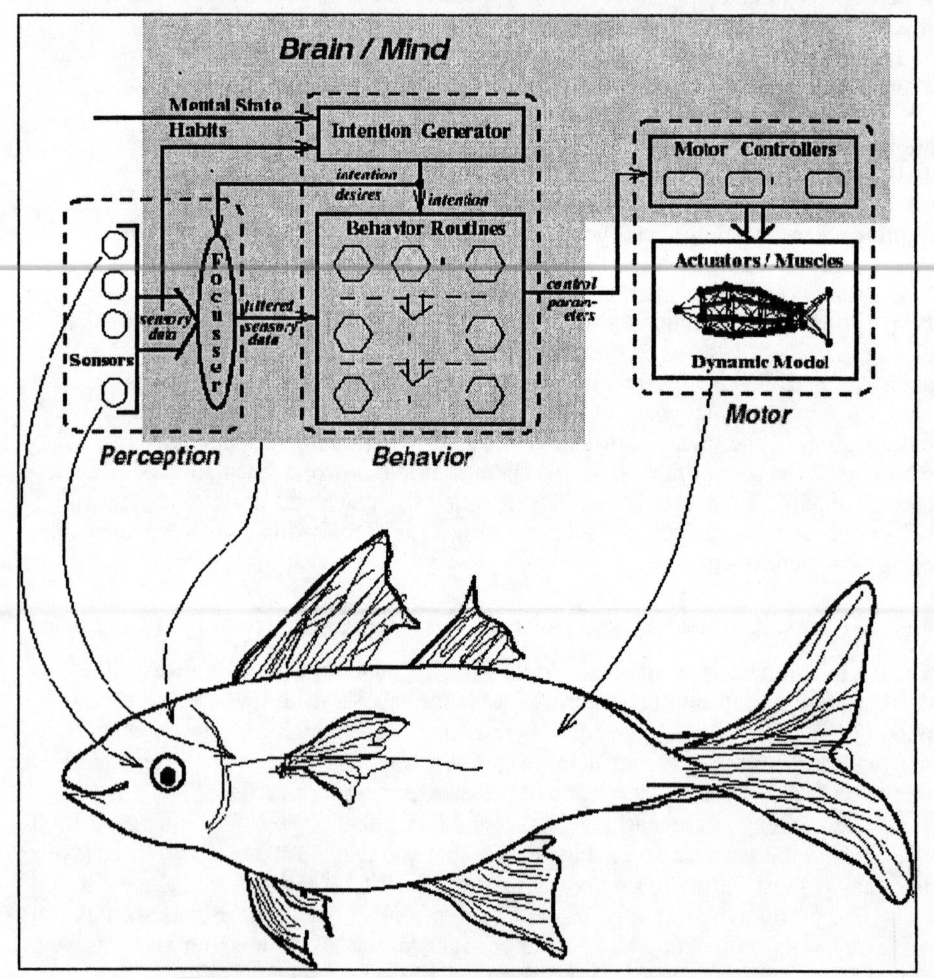

Figure 11. Diagram of an autonomous and avatar model fish.

The properties and internal control mechanisms of these artificial animals should be qualitatively similar to those of their natural counterparts. Specifically, we construct animal-like autonomy into traditional graphical models by integrating control structures for locomotion, perception and action. As a result, the artificial fishes and mammals have "eyes" and other sensors to actively perceive their dynamic environments; they have "brains" to interpret their perception and govern their actions. Just like real animals, they autonomously make decisions about what to do and how to do in everyday aquatic life, be it dramatic or mundane. This approach has been successfully demonstrated by the life-like virtual undersea world developed in [Tu 1996][Tu 1994].

We tackle the complexity of the artificial life model by decomposing it into three sub-models:

- *A graphical display model* that uses geometry and texture to capture the form and appearance of any specific species of sea animal.
- *A bio-mechanical model* that captures the physical and anatomical structure of the animal's body, including its muscle actuators, and simulates its deformation and dynamics.
- *A brain model* that is responsible for motor, perception and behavior control of the animal.

As an example of such an artificial life model, figure 11 shows a functional overview of the artificial fish that was implemented in [Tu 1996][Tu 1994]. As the figure illustrates, the body of the fish harbors its brain. The brain itself consists of three control centers: the motor center, the perception center, and the behavior center. These centers are part of the motor, perception, and behavior control systems of the artificial fish. The function of each of these systems will be previewed next.

6.5.4. Motor System

The motor system comprises the dynamic model of the sea creature, the actuators, and a set of motor controllers (MCs) which constitutes the motor control center in the artificial animal's brain. Since our goal is to animate an animal realistically and at reasonable computational cost, we seek to design a mechanical model that represents a good compromise between anatomical consistency, hence realism, and computational efficiency. The dynamic fish model [Blake 1983][Alexander 1992] represents a good example. It is important to realize that adequate model fidelity allows us to build motor controllers by gleaning information from bio-mechanical literature of the animal [Tinbergen 1951][Tansley 1965] Motor controllers are parameterized procedures, each of which is dedicated to carrying out a specific motor function, such as "swim forward", "turn left" or "ascend". They translate natural control parameters such as the forward speed, angle of the turn or angle of ascent into detailed muscle or fin/leg actions. Abstracting locomotion control into parameterized procedures enables behavior control to operate on the level of motor skills, rather than that of tedious individual muscle/limb movements. The repertoire of motor skills forms the foundation of the artificial animal's functionality.

Given that our application is in interactive VR systems, locomotion control and simulation must be executed at interactive speed. Physics-based dynamic models of animals offer physical realism but require numerical integration. This can be expensive when the models are complex.

There are various ways to speed things up. One way is to have multiple models of increasing complexity for each species of animal. For example, the simplest model will just be a particle with mass, velocity and acceleration (which still exhibits basic physical properties and hence can automatically react to external forces, such as water current, in a realistic manner). The speed-up can then be achieved by way of "motion culling", where when the animal is not visible or is far away, only the simplest model is used and the full mechanical model is used only when the animal is near by. This method, however, is view-dependent and can be tricky when the number of users is large. Another way is to use bio-mechanical models whenever possible (when they are simple enough to run the simulation at real-time) and with more complex animals, we can build pre-processed, parameterized motion libraries (which can be played back in real-time), like what is used by most of today's games. These (canned) motion libraries can be built from off-line simulations of elaborate bio-mechanical models of the animal or from motion-captured data. One thing to keep in mind is that, no matter what the underlying locomotion model is, its interface to the behavior system should be kept comparable as parameterized procedures of motor skills.

6.5.5. Perception System

Perception modeling is concerned with:

1. Simulating the physical and optical abilities and limitations of the animal's perception.
2. Interpreting sensory data by simulating the results of perceptual information processing within the brain of the animal.

When modeling perception for the purposes of interactive entertainment, our first task is to model the perceptual capabilities of the animal. Many animals employ eyes as their primary sense organ and perceptual information is extracted from retinal images. In a VR system, such "retinal" images correspond to the 2D projection of the 3D virtual world rendered from the point of view of the artificial animal's "eyes". However, many animals do not rely on vision as their primary perceptual mode, in which case vision models alone may not be able to appropriately capture the animal's perceptual abilities.

It is equally important to model the limitations of natural perception. Animal sensory organs cannot provide unlimited information about their habitats. Most animals cannot detect objects that are beyond a certain distance away and they usually can detect moving objects much better than static objects [Tansley 1965]. If these properties are not adequately modeled, unrealistic behaviors may result.

Moreover, at any moment in time, an animal receives a relatively large amount of sensory information to which its brain cannot attend all at once. Hence there must be some mechanism for deciding what particular information to attend to at any particular time. This process is often referred to as attention. The focus of attention is determined based upon the animal's behavioral needs and is a crucial part of perception that directly connects perception to behavior.

Unfortunately, it is not at all well understood how to model animal sensory organs, let alone the information processing in the brain that mediate an animal's perception of its world. Fortunately, for our purposes, an artificial animal in its virtual world can readily glean whatever sensory information is necessary to support life-like behavior by directly interrogating the world model and/or exploiting the graphics-rendering pipeline. In this way, our perception model synthesizes the results of perception in as simple, direct and efficient a manner as possible.

The perception system relies on a set of on-board virtual sensors to provide sensory information about the dynamic environment, including eyes that can produce time-varying retinal images of the environment. The brain's perception control center includes a perceptual attention mechanism which allows the artificial animal to train its sensors at the world in a task-specific way, hence filtering out sensory information superfluous to its current behavioral needs. For example, the artificial animal attends to sensory information about nearby food sources when foraging.

6.5.6. Behavior System

The behavior system of the artificial animal mediates between its perception system and its motor system. An *intention generator*, the animal's cognitive faculty, harnesses the dynamics of the perception-action cycle and controls action selection in the artificial animal. The animator establishes the innate character of the animal through a set of *habit parameters* that determine whether or not it likes darkness or whether it is a male or female, etc. Unlike the static habits of the animal, its mental state is dynamic and is modeled in the behavior system by several *mental state* variables. Each mental state variable represents a distinct desire. For example, the desire to drink or the desire to eat. In order to model an artificial animal's mental state, it is important to make certain that the modeled desires resemble the three fundamental properties of natural desires: (a) they should be time varying; (b) they should depend on either internal urge or external stimuli or both; (c) they should be satisfiable. The intention generator combines the habits and mental state with the incoming stream of sensory information to generate dynamic goals for the animal, such as to chase and feed on prey. It ensures that goals have some persistence by exploiting a single-item memory. The intention generator also controls the perceptual attention mechanism to filter out sensory information unnecessary to accomplishing the goal in hand. For example, if the intention is to eat food, then the artificial animal attends to sensory information related to nearby food sources. Moreover, at any given moment in time, there is only one intention or one active behavior in the artificial animal's behavior system. This hypothesis is commonly made by ethologists when analyzing the behavior of fishes, birds and four-legged animals of or below intermediate complexity (e.g. dogs, cats) [Tinbergen 1951][Manning 1979]. At every simulation time step, the intention generator activates behavior routines that input the filtered sensory information and compute the appropriate motor control parameters to carry the animal one step closer to fulfilling the current intention. The intention generator Primitive behavior routines, such as obstacle avoidance, and more sophisticated motivational behavior routines, such as mating, are the building blocks of the behavioral repertoire of the artificial animal.

6.6. Interfaces

Because it is our past experience that all technology much be bulletproof and withstand the rigorous demands of an unrelenting public, this equipment is designed along the lines of actual DIVE equipment which is very durable and rugged, big and easy to use.

6.6.1. Weight Belt with Dive Panel

This equipment enables visitors to interact with the environment and gain statistical information about the life species in the installation. The weight belt holds the battery pack and wireless communications, tracking and wand/pointer hardware, and a smart/memory card slot for recording individual data and experiences. Attached to it is a "Dive Panel" (cf. figure 12) which is a pointing wand, and small LCD monitor to display information about the species that are under investigation, and other climatic information. Through this, the visitor is able to access the knowledge base stored on the memory card and other data being generated real-time.

6.6.2. The FishGlove

The "FishGlove" enables visitors to control avatars throughout the reef and interact communicate with other avatars and autonomous life (cf. figure 13).

6.6.3. Enhanced Shutterglasses

Shutter glasses are connected to a timer so that when the visitor's use all their "air", they shut down and are inoperable. This makes the environment very difficult to see and forces the visitor to exit the installation. Some of the more sophisticated models for avatars and "Marine Biologists" have audio communications built in.

6.6.4. Taking Home a Bit of the Virtual Reef

Something else that we think is really important from NICE [Johnson 1998] is giving people some kind of artifact from the experience. Since VR is such a rare experience, and you really can't take anything physical away from the virtual space, it helps people when they try to describe what they did to other people, and enhance the experience and memory of the event.

Each participant in the virtual reef environment gets a smart/memory card to record a "narrative" of everything they did, characters they talked to, actions they performed. The narrative structure captures these interactions in the form of simple sentences such as: "Mary (disguised as a dolphin) stops by to see Murray the Moray eel to find out what he eats. Mary and Murray have dinner together".

The story sequence goes through a simple parser, which replaces some of the words with their iconic representations and stores the transcript onto their Memory card. This gives the story a "picturebook" style that the visitor can print out and keep to remember the various aquatic life they met that day.

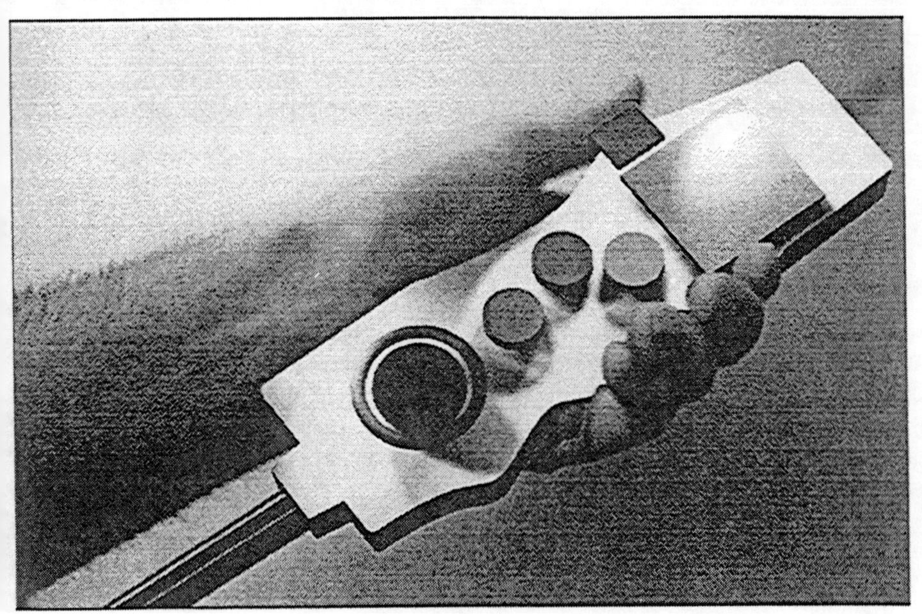

Figure 12. Dive Panel with pointing device and smart card access.

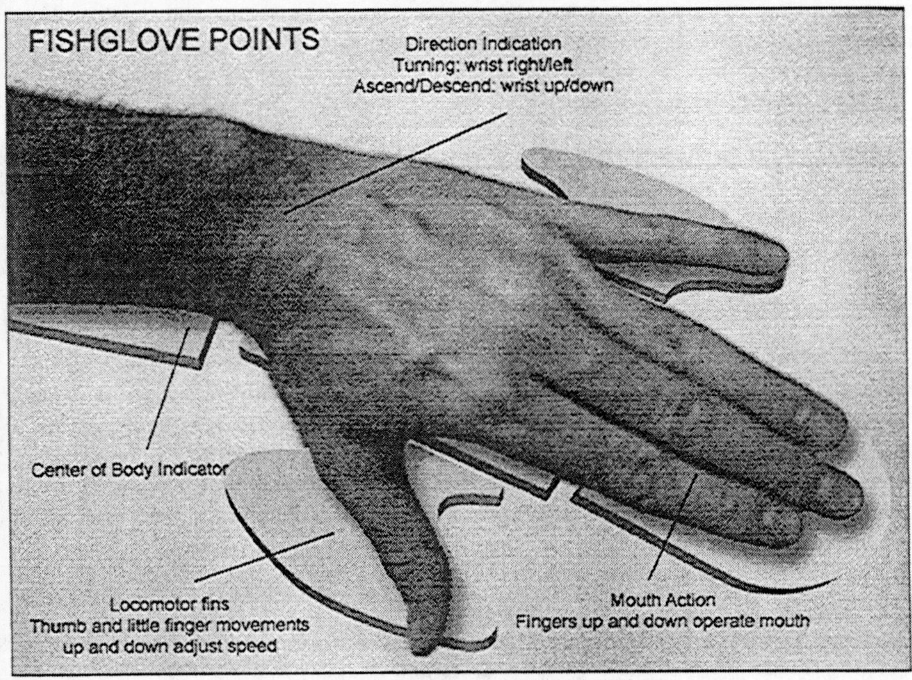

Figure 13. Control design of the FishGlove.

6.6.5. Smart/Memory Cards

We are currently investigating the use of Smart/Memory Cards to perform functions like: tracking and recording the user's actions, supplying a knowledge database, and providing financial calculations for visitation fees.

6.7. Interactivity

6.7.1. Visitor Interactivity

Avatar: There are many species in which a visitor can explore, from fish and other animals. Avatars can be as simple as the ones found at the localized exhibits for tourists to try out, to the sophisticated ones like dolphins and sharks. Avatars ultimately come under the domain of the simulated eco-cycle, so anything could happen: the visitor's avatar could be eaten at any time, and eventually will go through the entire life cycle of the reef.

Snorkeler: This level is simple roaming and exploring the environment with minimal hardware, such as shutter-glasses to immerse them in 3D. The smart card enables them to utilize the local interactive stands available throughout the environment.

Marine Biologist: This level of interaction is highly independent and is able to program their own interactivity levels for simulation, planning, studying, etc. The equipment this visitor has is the most advanced versions of the standard models.

6.7.2. Artificial Life Interactivity

The artificial life in the environment is just as interactive towards visitors as visitors are to them. They can and will interact with visitors in many different behaviors, from being friendly, territorial, viscous, funny, etc. Fast movement will scatter fish. In this way, true interactivity is two-way, creating a complex environment.

6.7.3. Staff Interactivity

Guides: These are sophisticated avatars that can interact with anyone in the environment. These people can be in the environment or hidden from view. They have audio and visual communication devices to drive avatars such as Mermaids, large fish, human animations or indigenous characters. This level of interactivity can be used for guided tours, assistants, observers and other special requirements. At this level, the guide is able to leave discovery objects which visitors can touch and find out more information.

6.8. Conclusion and Future Works

This system has been designed using existing and known solutions of virtual reality, networking, evolutional theory and artificial life, and yet, the innovation lies within the organization of each application into complexity. It has been designed with the forethought of being able to use the physical architecture as a generic shell to enable other stories besides the reef to be told. The project hosts many different components which when eventually compiled, will yield a first generation of a truly evolvable virtual reality environment.

6.9. Acknowledgements

The Gifu Research and Development Foundation and The International Society on Virtual Systems and MultiMedia, Gifu Japan.

The virtual reality research, collaborations, and outreach programs at the Electronic Visualization Laboratory (EVL) at the University of Illinois at Chicago are made possible through major funding from the National Science Foundation (NSF), the Defense Advanced Research Projects Agency, and the US Department of Energy; specifically NSF awards CDA-9303433, CDA-9512272, NCR-9712283, CDA-9720351, and the NSF ASC Partnerships for Advanced Computational Infrastructure (PACI) program. EVL also wishes to acknowledge Silicon Graphics, Inc. and Advanced Network and Services for their support. The CAVE and ImmersaDesk are trademarks of the Board of Trustees of the University of Illinois.

The Project Team:

Core Research Group:

Mr. Scot Thrane Refsland, VSL, Gifu University, JP - thrane@vsl.gifu-u.ac.jp
Prof. Andy Johnson, EVL, UIC, USA - ajohnson@eecs.uic.edu
Dr. Jason Leigh, EVL, UIC USA - jleigh@eecs.uic.edu
Mr. Carl Loeffler, Simation, PA, USA - cloeffler@aol.com
Dr. Xiaoyuan Tu, Intel Corp, USA - xiaoyuan.tu@intel.com

Complexity/Artificial Life/Evolvable Environments Team:

Prof. Victor DeLeon, CEC, FAU, USA - vic@laureate.cec.fau.edu
Dr. Claude Lattaud, Laboratoire d'Intelligence Artificielle de Paris-5, FR - latc@math-info.univ-paris5.fr
Mr. Glenn Proctor, British Telecom, UK, - gproctor@info.bt.co.uk

Underwater Camera, Projection 3D Vision Systems, Laser Projection Systems:

Dr. Ross Gilson, FAU, USA - ross@voyager.ee.fau.edu

Visuals/Design/Coordination:

Assoc. Prof. Jeff Jones, Com Design, QUT, AUS – ji.jones@qut.edu.au

Machine Vision Systems:

Prof. Sudhanshu Semwal, Unv Colorado/ATR JP - semwal@mic.atr.co.jp

Indigenous Content:

Mr. Brett Leavy, Indigenet, QANTM CMC, AUS - bleavy@qantm.com.au

Diving Coordination and Aquatic Equipment:

Craig Wilkins, Dive Master, Skipper, Cairns Australia

Organizations & World Heritage Authority Assistance:
International Society on Virtual Systems and MultiMedia, JP
UNESCO World Heritage Centre, FR
Great Barrier Reef Marine Park Authority, AUS

References

Alexander, R., 1992, *Exploring Bio-mechanics*, Scientific American Library (New York).

Blake, R., 1983, *Fish Locomotion*, Cambridge University Press (Cambridge).

Brooks, F.P., Ouh-Young, M.J., Batter, J.J., Kilpatrick, P.J., 1990, Project GROPE - Haptic Displays for Scientific Visualization, *Proceedings of SIGGRAPH'90*, **24**, 177.

Capin, T.K., Noser, H., Thalmann, D., Pandzic, I.S., Thalmann, N.M., 1997, Virtual Human Representation and Communication in VLNet Networked Virtual Environment, *IEEE Computer Graphics and Applications*, **17**, 42.

Cruz-Neira, C., Sandin, D.J., and DeFanti, T.A. 1993, Surround-Screen Projection-Based Virtual Reality: The Design and Implementation of the CAVE, *Proceedings of ACM SIGGRAPH '93 Computer Graphics Conference*, 135.

Eric, W., Grimson, L., 1990, *Object Recognition by Computer: The Role of Geometric Constraints*, MIT Press (Cambridge).

Faugeras, O., 1996, *Three-dimensional Computer Vision: A geometric Viewpoint*, MIT Press (Cambridge).

Geiger, D., Liu, T.L., 1996, Recognizing Articulated Objects with Information Theoretic Methods, *Proceedings of International Conference on Automatic Face-and-gesture Recognition* (Killington), 45.

Goldberg, D., 1989, *Genetic Algorithms in Search, Optimization and Machine Learning*, Addison Wesley.

Gottschalk, S., Hughes, J.F., 1993, Autocalibration for Virtual Environments Tracking Hardware, *Proceedings of SIGGRAPH 1993*, 65.

Holland, J., 1992, *Adaptation in Natural and Artificial Systems – 2nd edition*, MIT Press.

Johnson, A., Roussos, M., Leigh, J., Barnes, C., Vasilakis, C., Moher, T., 1998a, The NICE Project: Learning Together in a Virtual World, *Proceedings of VRAIS '98* (Georgia), 176.

Johnson, A., Leigh, J., Costigan, C., 1998b, Multiway Tele-Immersion at Supercomputing '97, To appear in *IEEE Computer Graphics and Applications*.

Krueger, M.W., 1991, *Artificial Reality II*, Addison Wesley Publishing Company (Reading), MA.

Lattaud, C., 1998, A Macro Mutation Operator in Genetic Algorithms, *Proceedings of the 13th European Conference on Artificial Intelligence*, edited by Henri Prade, Wiley (Brigthon).

Leigh, J., DeFanti, T., Johnson, A., Brown, M., Sandin, D., 1997, Global Tele-Immersion: Better than Being There, *Proceedings of 7th International Conference on Artificial Reality and Tele-Existence* (Tokyo), 10.

Lorenz, E.N., 1963, *Journal of Atomic Sciences*, **20**, 130.

Manning, A., 1979, *An Introduction to Animal Behavior – 3rd edition*, Addison-Wesley Publications.

Maybank, S., 1993, Theory of Reconstruction from Image Motion, Springer-Verlag.

Meyer, K., Applewhite, H.L., Biocca, F.A., 1992, A Survey of Position Trackers, *Presence*, MIT Press, **1**, 173.

Piaget, J., 1973, *To Understand is to Invent: The Future of Education*, Grossman (New York).

Rashid, R.F., 1980, Towards a system for the Interpretation of Moving Light Displays, *IEEE Transaction on PAMI*, **2**, 574.

Robotix, *Robotix Mars Mission*, Carnegie Science Center, http://www.csc.clpgh.org/Exhibits/Mars/index.htm.

Semwal, S.K., Hightower, R., Stansfield, S., 1996, Closed form and Geometric Algorithms for Real-Time Control of an Avatar, *Proceedings of IEEE VRAIS'96*, 177.

Semwal, S.K., Ohya, J., 1998, The Scan&Track Virtual Environments, *Virtual Worlds*, edited by J.C. Heudin, Proceedings of the 1st International Conference on Virtual Worlds, Springer LNCAI, 1434, 63.

Serra, J., 1988, *Image Analysis and Mathematical Morphology*, Academic Press, **1** & **2**.

State, A., Hirota, G., Chen, D.T., Garrett, W.F., Livingston, M.A., 1996, Superior Augmented Reality Registration by Integrating Landmark Tracking and Magnetic Tracking, *Proceedings of SIGGRAPH'96*, 429.

Sturman, D.J., Zeltzer, D., 1994, A Survey of Glove-based Input, *IEEE CG&A*, 30.

Sutherland, I.E., 1968, A head-mounted three-dimensional display, *Proceedings of ACM Joint Computer Conference*, **33**, 757.

Tansley, K., 1965, Vision in Vertebrates, Chapman and Hall (London).

Tinbergen, N., 1951, The Study of Instinct, Clarendon Press (Oxford).

Talbert, N., 1997, Toward Human-Centered Systems, *IEEE CG&A*, **17**, 21.

Tu, X., Terzopoulos, D., 1994, Artificial fishes: Physics, locomotion, perception, behavior, *Computer Graphics*, Annual Conference Series, Proceedings ACM SIGGRAPH'94 (Orlando), 43.

Tu, X., 1996, Ph.D Thesis, Department of Computer Science, University of Toronto. http://www.dgp.toronto.edu/people/tu.

Virtuarium, *Goto Virtuarium DOME system*, www.bekkoame.or.jp/~goto-co/.

Wejchert, J., Haumann, D., 1991, Animation aerodynamics, *Computer Graphics*, ACM SIGGRAPH'91, **25**, 19.

Wren, C., Azarbayejani, A., Darrell, T., Pentland, A., 1996, Pfinder: Real-Time Tracking of the Human Body, *International Conference on Automatic Face and Gesture Recognition*, 51.

Chapter 7

Virtual Worlds and Complex Systems

Yaneer Bar-Yam
New England Complex Systems Institute
Yaneer@necsi.org
http://necsi.org

7.1. Abstract

The scientific study of complex systems can benefit from the use of computer based virtual representations that capture experimental observations of real systems and test theoretical knowledge. The development of virtual environments can benefit from the understanding and specific knowledge obtained in the study of complex systems as it relates to the properties of physical, biological and social systems and their environments. The meeting of these two efforts in the context of virtual worlds is an exciting opportunity for complementarity and synergy. In this paper we discuss the use of the complexity profile to quantify the amount of information needed to describe a system as a function of the scale/precision of observation. Using the complexity profile it is possible to characterize the properties of simple and complex systems and to guide the development of computer based representations of these systems.

7.2. Introduction

7.2.1. Representations

The construction of virtual environments for entertainment, educational and other purposes, has gained interest in recent years based upon improvements in computer based graphics through high speed processing and rendering. However, this endeavor has also promoted the development of a better understanding of useful and effective methods for representing realistic or imagined worlds. The scientific study of complex systems has also gained interest in recent years. This field is devoted to understanding the general properties of systems and their environments. Understanding the relationship between systems and their representations is central to this discipline. The construction of artificial worlds merges these two fields providing an opportunity to explore the nature of representations that capture aspects or attributes of the real world around us or the possibilities inherent in worlds that differ in minor or major aspects from the real one. In a sense the development of virtual worlds is a necessary extension of scientific progress-as the level of our understanding deepens, the only way to test this understanding is to build representations which can be compared with observed properties of the world.

For example, How can we know that we understand the properties of a biological organism without constructing a representation of it which can serve to replicate its behavior? Similarly, if we want to understand social or historical processes we must demonstrate this understanding by creating artificial representations which demonstrate this understanding. This argument about the necessity of artificial representations to demonstrate understanding is complementary to the notion that knowledge is useful when implemented in artificial constructs. It is also complementary to the notion that artificial or virtual worlds can explore the realm of the possible, not only what is found around us.

Many of the underlying concepts in the study of complex systems [Kauffman 1995][Holland 1995][Bak 1996][Simon 1996][Bar-Yam 1997] are relevant to the formation of virtual worlds. Among these underlying concepts are the notions of concise representation by algorithmic compression, and the establishment of laws that lead to consistent patterns of behavior within an artificial world. Somewhat less well appreciated are other principles that underlie complex systems: the existence of substructure, the importance of composites in the creation of new organisms or patterns of behavior, the existence of various time scales, and the necessity for a systematic relationships between the behavior of individual elements of the worlds and their surroundings.

7.2.2. Precision, Redundancy and Information

One of the issues which is central to representations is understanding how precise a representation can or should be in order to capture reality or to capture the aspects of reality that are intended. To think about these issues it is helpful to consider the problem of systematically developing more and more precise representations of the world or some smaller system.

Intuitively we can understand that more and more precise representations require progressively more information to describe. A standard approach to progressive representation might be based upon a pixel-by-pixel representation. Such a representation is a digitized movie representing the space-time of a system either as a two+one dimensional view (standard movie) or a three+one dimensional history of the system which includes internal structure of objects in the movie.

Starting from a particular representation, we could obtain a rougher, less precise, representation by local averaging of pixels to obtain fewer pixels and fewer frames per second. This is a standard approach to reducing the length of computer representations of video or still images to allow for easier storage or communication.

Restricting ourselves to a pixel based representation we can create a series of progressively coarser representations which capture progressively less detail. To completely specify the representation we should also describe the number of bits per pixel that are used. As we coarsen the representation we might keep the number of bits per pixel the same, reduce the number of bits per pixel or even increase the number of bits per pixel. While it is not necessarily obvious, even in the latter case there are effectively fewer bits in the coarser representation. Thus, using local averaging we can construct a hierarchy of representations with systematically fewer bits in the representation as a whole.

This notion of a set of systematically coarser representations has many practical uses. Indeed, we can readily understand that by thinking about an observer who sees the system from farther and farther away, the system will appear to the observer to be realistic at coarser and coarser representations. The notion of a "realistic" representation must be carefully considered. A photograph of a person from several miles away can be easily represented realistically. What is important is that there is no fundamental value of representation precision which is "correct". The level of precision required depends on the observer, or, more correctly, the relationship between the observer and the system being observed.

This idea has been recognized in more traditional forms of virtual reality by painters and sculptors[1], who have explored various mechanisms to increase the realism of their art. Realistic paintings always rely upon the essential properties of the observer, which includes also the intended viewing distance and perspective.

A practical application of the notion of the precision of representation in traditional art is the purposeful absence of details in an image. This increases realism by suggesting or imposing a limit of detail which is possible for the observer to see. The level of detail may be varied with "distance" of the virtual image as a whole or differential levels of details of the parts. Without other clues, the observer interprets this as defining the distance to the parts of the image rather than as a property of the image itself.

A more sophisticated perspective on the development of progressively finer levels of observation requires us to incorporate the notion of algorithmic representations of patterns in the system that is represented. Thus, we can allow the wide range of compression algorithms that take the pixel based image and represent redundancies, patterns or systematic features of the image in a more concise way. The existence of patterns and features implies the existence of redundancy. By eliminating redundancy a more concise representation can be constructed. However, it is important to recognize that the most concise representation for observation at a particular scale is always longer than (or equal to) the most concise representation at a coarser scale. This follows because we could simply replace the representation at the coarser scale by the representation at the finer scale if it ware not shorter.

[1] I thank Sue Wilcox for pointing out this connection.

7.2.3. The Concept of a Complexity Profile

A more systematic understanding of representations as a function of scale can be introduced using the "complexity profile". The complexity profile is a measure of the minimum amount of information (bits) needed to represent a system as a function of the degree of precision of the description. The importance of this quantity for the practical development of artificial worlds is that it informs us about the level of effort and the amount of storage that is needed in order to reach a certain degree of precision. Alternatively, if we are limited to a certain amount of information, the complexity profile informs us of the level of precision that can be obtained.

While the complexity profile can be thought about as the amount of information necessary to describe a system as a function of scale, it is also possible to think about the complexity profile as characterizing the relationship between the behavior of parts of the system and the behavior of the system as a whole. This interpretation arises when we recognize that redundancy arises from correlations between parts of the image. Thus we can think about scale as reflecting the degree of correlation in the behavior of the parts of the system rather than as the size of pixels. This approach is central to the general study of complex systems which strives to understand how the behavior of the parts gives rise to collective behaviors.

The complexity profile is a means of focusing attention on the scale at which a certain behavior of a system is visible to an observer. The complexity profile counts the number of independent behaviors that are visible at a particular scale, and includes all of the behaviors that are visible at larger scales. With the complexity profile, one can explicitly relate the scale of behaviors to the independence or interdependence of system components. Focusing on the scale of behaviors enables us to consider the relationship between behaviors on different scales, and between parts and wholes.

The general properties of the complexity profile are illustrated by the examples of independent and coherent behavior. Consider a system in which the parts behave independently. The system behavior at a small scale requires specifying what each of the parts is doing. However, when observing on a larger scale, it is not possible to distinguish the individual parts. Thus, the system behavior is simple. Contrast this independent behavior with coherent motion. In coherent motion all of the parts of the system move in the same direction. Since the behaviors of the parts are all the same, they are simple to describe on the largest scale. Moreover, once the largest scale behavior is described, the behavior of each of the parts is also known.

Neither of these two examples corresponds to complex collective behavior. Unlike the coherent motion case, complex behavior must include many different behaviors. Unlike the independent action case, many of these behaviors must be visible on a large scale. In order for such visibility to occur, various subgroups of the system must have coordinated behaviors. The resulting behaviors are distributed at different scales and the complexity profile decreases gradually with increasing scale.

These concepts can be applied to physical, biological and social systems. In the following sections we expand the explanation of scale and the complexity profile to illustrate features that should help their diverse application to simple and complex systems that we wish to understand or represent.

7.3. Independence, Coherence and the Scale of Observation

All macroscopic systems, whether their behavior is simple or complex, are formed out of a large number of parts. The following examples suggest insights into how and in what way simple or complex behaviors arise.

Inanimate objects generally do not have complex behaviors. Notable exceptions include water flowing in a stream or boiling in a pot, and the atmospheric dynamics of weather. However, if water or air are not subject to external force or heat variations, their behavior is simple. Nevertheless, by looking very closely, it is possible to see the rapid and random thermal motion of atoms. Describing the motion of all of the atoms in a cubic centimeter of water would require a volume of writing which is more than ten billion times the number of books in the Library of Congress. Though this would be a remarkably large amount of information, it is all irrelevant to the visible macroscopic behavior of a cup of water.

Thermodynamics and statistical mechanics explained this paradox at the end of the 19th century. The generally independent and random motion of atoms means that small regions of equal size contain essentially the same number of atoms. At any time the number of atoms leaving a region and the number of atoms entering it are also essentially the same. Thus, the water is uniform and unchanging.

While biological organisms generally behave in a more complex way than inanimate objects, independent and randomly moving biological microorganisms also have simple collective behavior. Consider the behavior of microorganisms that cause diseases. What is the difference between the microorganisms and the cells that form a human being? From a macroscopic perspective, the primary difference is that a large collection of microorganisms do not result in complex collective behavior. Each of the microorganisms follows an essentially independent course. The independence of their microscopic actions results in an average behavior on a large scale which is simple. This is true even though, like the human being, all of the microorganisms may originate from a single cell.

There is a way in which the microorganisms do act in a coherent way-they damage or consume the cells of the body they are in. This coherent action is what enables them to have an impact on a large scale. It is only because many of them perform this action together that makes them relevant to human health.

The notion of coherence also applies to physical systems. Atoms at room temperature in a gas, liquid or solid, move randomly at speeds of 1000 km/hr but have less large scale impact than an object thrown at much slower speed of 50 km/hr. It is the collective coherent motion of all of the atoms in the object that enables them to have impact on a large scale.

Thus, there are two paradigms for simple collective behavior. When the parts of a system have behaviors that are independent of each other, the collective behavior of the system is simple. Close observation reveals complex behavior of the parts, but this behavior is irrelevant to the collective behavior. On the other hand if all parts act in exactly the same way, then their collective behavior is simple even though it is visible on a very large scale.

These examples of behavior can also be seen in the social systems. We can include a simple analysis of the historical progression of human civilization.

Primitive tribal or agrarian cultures involved largely independent individuals or small groups. Military systems involved large coherent motions of many individuals performing similar and relatively simple actions. These coherent actions enabled impact at a scale much larger than the size of the military force itself.

By contrast, civilization today involves diverse and specialized individual behaviors that are nevertheless coordinated. This specialization and coordination allow for highly complex collective behaviors capable of influencing the environment on many scales. Thus the collective behavior of human civilization arises from the coordinated behavior of many individuals in various groupings.

7.4. The Complexity profile

It is much easier to think about the problem of understanding collective behavior using the concept of a complexity profile. The complexity profile focuses attention on the scale at which a certain behavior of a system is visible to an observer, or the extent of the impact it can have on its environment. Both of these are relevant to interactions of a system with its environment-an observer can see the behavior only when the behavior is sufficiently large to affect the observer.

A formal definition of scale considers the spatial extent, time duration, momentum and energy of a behavior. More intuitively, when many parts of a system act together to make a single behavior happen, that behavior is on a large scale, and when few parts of a system act together, that behavior is on a small scale. The energy of different actions of the system is also relevant. When the amount of energy devoted to an action is large, then it is a large scale action. In essence, the units of energy are working together to make a large scale behavior. A more systematic treatment of the scale of particular behaviors leads to the complexity profile.

The complexity profile counts the number of independent behaviors that are visible at a particular scale and includes all of the behaviors that have impact at larger scales. The use of the term "complexity" reflects a quantitative theory of the degree of difficulty of describing a system's behavior. In its most basic form, this theory simply counts the number of independent behaviors as a measure of the complexity of a system. The complexity profile characterizes the system behavior by describing the complexity as a function of scale.

The central point is: When the independence of the components is reduced, scale of behavior is increased. To make a large collective behavior, the individual parts that make up this behavior must be correlated and not independent. This reduction of independence means that describing the collective behavior includes part or all of the behavior of the parts and therefore our description of the parts is simpler. When the behaviors of parts are coupled in subgroups, their behavior is manifest at the scale corresponding to the size of the group.

Thus, fixing the material composition and the energy of the system, there are various ways the system can be organized. Each way of organizing the system and distributing the energy through the system results in tradeoffs between the complexity of their microscopic description against the complexity of their description at progressively larger scales.

To illustrate the complexity profile, consider a system in which the parts behave independently. The system behavior at a small scale requires specifying what each of the parts is doing. However, when observing on a larger scale, it is not possible to distinguish the individual parts even in a small region of the system, only the aggregate effect of their behavior is observable. Since their behaviors are independent, they cancel each other in their impact on the environment. Thus, the description of the system behavior is simple. The behavior of each individual part disappears upon averaging the behavior of the local group. Examples of this include microorganisms swimming randomly in a pond or people moving around in a crowd that does not move as a whole. When one person goes one way, another person fills his place and together there is no collective movement.

Independent behavior is to be contrasted with coherent motion. In coherent motion all of the parts of the system move in the same direction. This is the largest scale behavior possible for the system. Since the behaviors of the parts of the system are all the same, they are simple to describe on the largest scale. Moreover, once the largest scale behavior is described, the behavior of each of the parts is also known.

Neither of these two examples corresponds to complex collective behavior. Unlike the coherent motion case, complex behavior must include many different behaviors. Unlike the independent action case, many of these behaviors are visible on a large scale. In order for such visibility to occur various subgroups of the system must have coordinated behaviors. The resulting dynamic correlations are distributed at different scales. Some of them are found at a microscopic scale in the coupled motion or positions of molecules, and others appear in the collective motion of, for example, muscle cells and the motion of the body as a whole. Thus, the complexity profile of a complex system like a human being involves a distribution of scales at which behavior manifests itself. This balance between highly random and highly ordered motion is characteristic of the behavior of complex systems.

The discussion of independent, coherent and complex behavior can be applied to physical, biological or social systems. Think about the gas molecules that bounce independently in a room, or the coherent alignment of magnetic regions of a magnet. In the former case, all of the parts of the system act independently and the complexity profile resembles the independent component example. In the latter, the parts of the system are all aligned, and there is a large scale behavior.

In biological systems a collection of microorganisms may act essentially independently, and a disease microorganism by multiplying and acting coherently in attacking the human body can have impact on a much larger scale. Finally, the cells of the body are interdependent and have collective complex behavior.

In social systems a crowd of people may move aimlessly with the random motions of individuals canceling each other so that the crowd does not move significantly. By contrast, an army is trained to move coherently over large distances and long times resulting in a large scale impact. Complex behaviors are realized by human organizations such as corporations that can have behaviors distributed over many scales.

7.5. Conclusions: Application to Virtual worlds

In order for virtual worlds to capture reality they must represent or capture the fundamental properties of the world. Physical laws such as gravity and the trajectories of objects represent only a limited view of the impact of laws on the behavior of complex systems. When we consider the complexity profile, we recognize that a realistic portrayal of the world must have a well defined scale of observation. Providing fantastic detail in one aspect and poor detail in another does not give a sense of reality. This notion, already captured in the world of art, should be an important aspect of the development of virtual worlds as both the static structure and the dynamic properties of images are improved.

Finally, in the development of virtual worlds which do not strive to represent the real world, but some imagined place, it is still relevant to consider the level of precision of representation and how this affects the observer's understanding of the world.

References

Bak, P., 1996, *How Nature Works: The Science of Self-organized Criticality*, Copernicus, Srpinger Verlag (New York).

Bar-Yam, Y., 1997, *Dynamics of Complex Systems*, Addison-Wesley (Reading).

Holland, J.H., 1995, *Hidden Order: How Adaptation Builds Complexity*, Helix Books, Addison-Wesley (Reading).

Kauffman, S.A., 1995, *At Home in the Universe*, Oxford University Press (New York).

Simon, H.A., 1996, *The Sciences of the Artificial*, 3rd edition, MIT Press (Cambridge).

Chapter 8

Investigating the Complex with Virtual Soccer

Itsuki Noda
Ian Frank
Complex Games Lab
Electrotechnical Laboratory
noda, ianf@etl.go.jp

8.1. Introduction

The rules of soccer, or association football, were formulated in 1863 in the UK by the Football Association, and the game has since become one of the most widely played in the world. The popularity of the game is illustrated by the following passage:

> *Contrary to popular belief, the world's greatest sporting event in terms of prolonged, worldwide interest is not the Olympic Games. Rather, it is the World Cup of football, which, like the Olympics, is held just once every four years and is played out over a period of two weeks or more. The United States hosted this global spectacle in 1994– and what a spectacle it turned out to be. The championship game between Brazil and Italy was witnessed live by a crowd of over 100,000 people in the Rose Bowl in Pasadena, California, and by a crowd of at least a billion on television the world over.*

We quote this description at the start of this chapter not simply because it illustrates the popularity of soccer, or because the original version of this paper was presented at the same time as France-98, the successor to the American World Cup. Rather, we are just as interested in the source of the excerpt – not a sports book or newspaper, but the opening paragraph of the first page of a book on science: *Would-be Worlds*, by John Casti [Casti 1997a].

Casti's purpose in describing the football World Cup is to introduce the notion of *complex systems*. Football is an excellent example in this context because it is not possible to understand the game by analyzing just individual players. Rather, it is the interactions in the game, both between the players themselves and between the players and the environment, that determine the outcome. This is the essence of complex systems: they resist analysis by decomposition.

We share Casti's belief that the arrival of cheap, powerful widespread computing capability over the past decade or so has provided us with a tool for studying complex systems as complete entities. Casti himself suggests that such computer simulations (virtual worlds, would-be-worlds) "play the role of laboratories for complex systems", which, in contrast to more conventional science laboratories "allow us to explore information instead of matter" [Casti 1997b]. It is precisely to provide this kind of laboratory for research on complex systems that we have developed a simulation of the game of soccer. This system, Soccer Server, enables a soccer match to be played between two teams of player-programs. Soccer Server provides a virtual soccer field (*e.g.*, see Figure1) and simulates the movements of players and a ball. A client program can then provide the "brain" of a player by connecting to the Soccer Server via a computer network (using a UDP/IP socket) and specifying actions for that player to carry out. In return, the client receives information from the player's sensors. In July of 1997, Soccer Server was used to stage the simulation league of the First Robot Soccer World Cup (RoboCup), in Nagoya, Japan. Twenty-nine teams by researchers from ten different countries competed in this competition. In 1998, an updated version of Soccer Server was used to stage the second RoboCup in Paris. This contest, held just before the semi-finals of France-98, increased the number of teams to 36 and the number of participating countries to thirteen.

In the following pages, we describe Soccer Server, identifying in particular why soccer is such a suitable choice as a simulation domain. We also take care to evaluate the success of Soccer Server, demonstrating that the properties of *fidelity*, *simplicity*, *clarity*, *bias-free*, and *tractability* are effective measures for assessing this kind of virtual world. Finally, we introduce and discuss the notion of *focussing*, describing how Soccer Server is being used to investigate the most effective ways of visualizing real-time complex systems.

8.2. Soccer as a Domain for Modeling

There were many reasons behind the choice of soccer as a simulation domain. Initially, of course, there was the appeal that soccer is fun and easily understood by large numbers of people. More important than this, however, was the significant challenge offered by trying to formalize and understand the domain itself.

8.2.1. Soccer as a Complex System

To illustrate the difficulty of the soccer domain, let us return to the discussion of complex systems that we started in the Introduction, and to Casti's book, *Would-be Worlds*. In discussing how a theory of the complex might be formalized, Casti introduces a number of characteristics he describes as the "key components" of complex, adaptive systems. We reproduce this list here [Casti 1997a]:

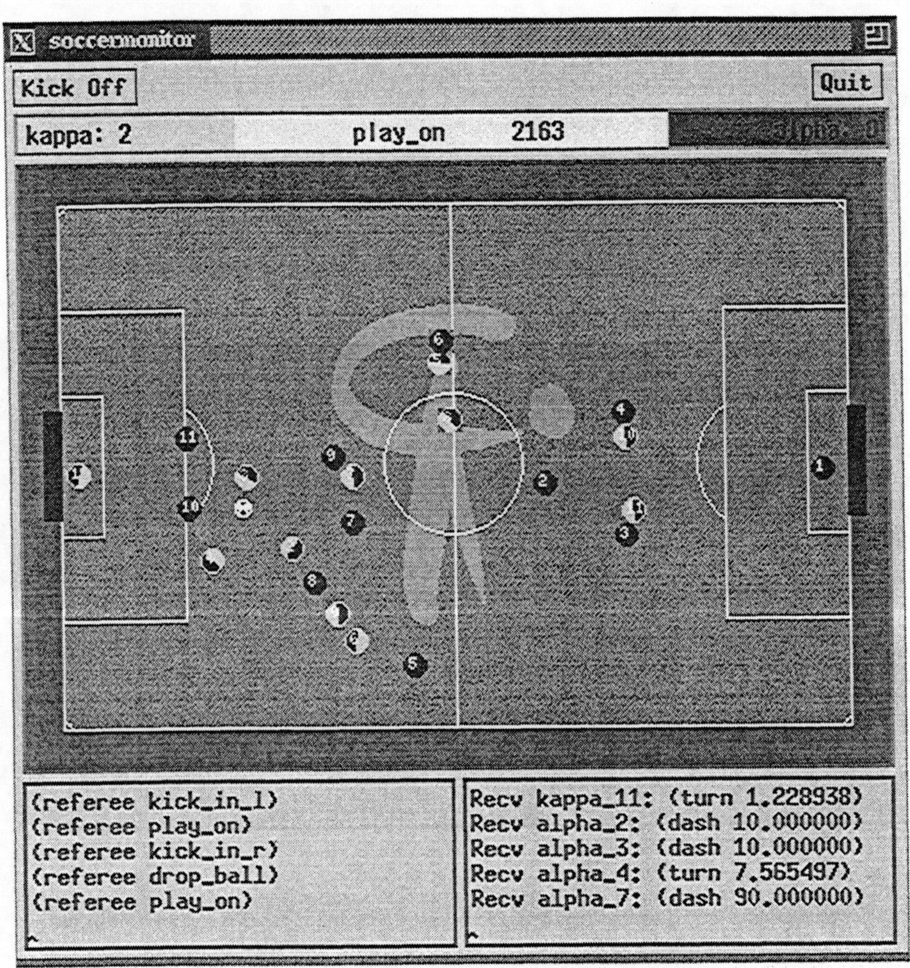

Figure. 1. Window image of Soccer Server.

• *Medium-sized number of agents*: Complex systems have a number of agents that is not too small and not too big, but just right to cause interesting patterns of behavior.

• *Intelligent and adaptive*: Not only are there a medium number of agents, these agents are intelligent and adaptive. This means that they make decisions on the basis of rules, and that they are ready to modify the rules they use on the basis of new information that becomes available. Moreover, the agents are able to generate new rules that have never before been used, rather than being hemmed in by having to choose from a set of preselected rules for action.

• *Local information*: No single agent has access to what *all* the other agents are doing. At most each agent gets information from a relatively small subset of the set of agents and processes this "local" information to come to a decision as to how he or she will act.

It should not be too hard to convince the reader that the game of soccer is a very good fit for these properties. Soccer has 22 agents. This falls comfortably between Casti's examples of systems with large numbers of agents (galaxies, which have enough agents to be treated statistically), and small numbers of agents (conflicts between two superpowers). The agents in soccer are also intelligent and adaptive, with the players on each team striving to perform well together, and to out-manœuvre the opponents. Finally, the information in soccer is clearly limited, as the players can only see in the direction they are facing, and verbal communication is hampered both by distance and by the importance of concealing intentions from the opponents.

Thus, understanding soccer is a significant challenge: it has the characteristics of a complex system, for which, as Casti emphasizes, "there exists no decent mathematical theory" [Casti 1997b]. However, as well as being a good example of a complex system, another important factor also influenced our choice of soccer: the feasibility of constructing a simulation. We examine this question below.

8.2.2. The Nature of Play

One of the foremost tasks in the design of a simulation is the creation of a *model* of the domain in question; it is the implementation of the rules and principles of the model that produces the simulation. In this respect, we found that soccer was a very amenable domain, simply by virtue of being a game.

An excellent study of the nature of play is given by Huizinga [Huizinga 1950]. Although primarily interested in the significance of play as a cultural phenomenon, some of the Huizinga's explanations are relevant to our theme here. Of particular interest is Huizinga's identification of a small number of main characteristics of play. For instance, the first two properties of play identified by Huizinga are that it is voluntary (or, as Huizinga says, "Play to order is no longer play"), and that play is not "ordinary" or "real" life. These properties already suggest some kind of modeling process. Huizinga then notes that "all play is limited in locality and duration" – another feature that makes play a promising candidate for convenient simulation.

Further general properties identified by Huizinga are that "all play has binding rules" and that "play can be repeated and transmitted". The presence of rules means that part of the work of modeling is already done. Indeed, in implementing Soccer Server, we took the simple approach of basing our model primarily on just the rules of the game. The remainder of the Soccer Server model was then formulated to allow maximum scope for the investigation of how the rules of soccer can actually be followed in the best way; that is, how to create a team that can play well. The property of repeatability and transmittability then indicates how our model can be used as an ideal testbed for allowing researchers to investigate and discuss the properties of the domain.

8.2.3. The Type of Model Represented by Soccer Server

As we suggested above, Soccer Server models the rules of the game of soccer in a way that allows the system to be used as a testbed for research on the nature of the domain. The intention is that different soccer player-control algorithms can be tested against each other to discover which are stronger. Thus, Soccer Server can be viewed as a predictive model of the relative strengths of these algorithms. We agree with [Casti 1997a] that "the first and foremost test that a model must pass is that it must provide convincing answers to the questions we put to it". Note, though, that the questions we want to answer with Soccer Server are not just which algorithms perform better, but also *why* the better algorithms are superior. This is a consequence of viewing the system as a testbed for research; the ultimate goal is to advance the theory of complex systems.

We should also note that although we concentrate here on the virtual world created by Soccer Server, the RoboCup tournaments themselves also include competitions for teams of real robots. In our view, this link is very important, since there are many results that show that real-world experiments can produce results that could not be expected from simulation alone [Thompson 1997][Brooks 1986]. With this caveat said, then, let us move on to take a detailed look at our implementation of a model of soccer.

8.3. Soccer Server Itself

Soccer Server enables a soccer match to be played between two teams of player-programs (possibly implemented in different programming systems). A match using Soccer Server is controlled using a form of client-server communication. Soccer Server provides a virtual soccer field (such as the one we presented in figure 1) and simulates the movements of players and a ball. A client program can provide the "brain" of a player by connecting to the Soccer Server via a computer network (using a UDP/IP socket) and specifying actions for that player to carry out. In return, the client receives information from the player's sensors.

Figure 2 gives an overview of how the Soccer Server communicates with clients. The three main modules in the Soccer Server itself are:

1. *A field simulator module*. This creates the basic virtual world of the soccer field, and calculates the movements of objects, checking for collisions.

2. *A referee module*. This ensures that the rules of the game are followed.

3. *A message-board module*. This manages the communication between the client programs.

Below, we describe each of these modules in turn. We also describe how uncertainty is represented in the world of the Soccer Server. Note that a large collection of related material, including sources and full manuals, is maintained at the Soccer Server home page: http://ci.etl.go.jp/ noda/soccer/server.

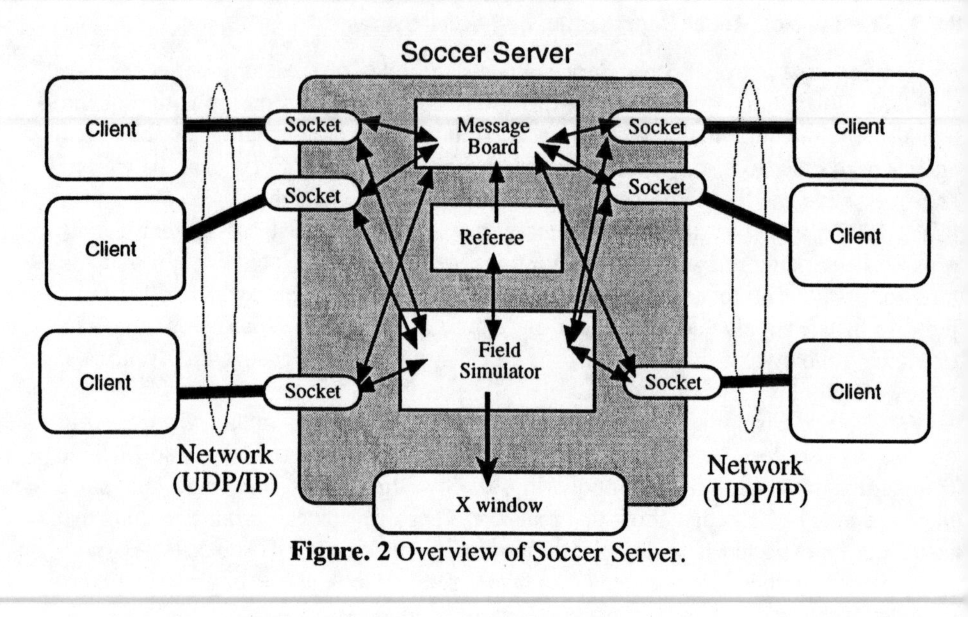

Figure. 2 Overview of Soccer Server.

8.3.1. Simulator Module: The Basic Virtual World

The soccer field and all objects on it are 2-dimensional. This means that, unlike in real soccer, the ball cannot be kicked in the air and the players cannot make use of skills such as heading or volleying. The size of the field, measured in the notional internal equivalent of meters, is 105m x 68m. On the field, there are objects that move (the players and the ball) and objects that are stationary (flags, lines and goals). The stationary objects serve as landmarks that clients can make use of to determine their player's position, orientation or speed and direction of movement. Clients are not given information on the absolute positions of their own players, but only on the relative location of other objects. Thus, the fixed objects are important reference points for the client programs' calculations.

The moving objects (the players and the ball) are treated as circles, as shown in the screen capture of Figure 3. The simulation of their motion is carried out object by object in the following stepwise fashion: (1) if the object is a player, any changes in the player's velocity (or the velocity of the ball) resulting from commands issued by the client controlling that player are calculated, (2) the velocity of the object is added to its position, and (3) the object's velocity decays at a pre-determined rate. Collisions are also simplified. If, after being moved, one object overlaps a second (that is, two objects collide with each other), the first object is moved back until the overlap is removed and its velocity is multiplied by -0.1.

All communication between the server and each client is in the form of ASCII strings. Therefore, clients can be realized in any programming environment on any architecture that has the facilities of UDP/IP sockets and string manipulation. The protocol of the communication consists of:

• *Control commands*: messages sent from a client to the server to control actions of the client's player. The basic commands are turn, dash and kick. Communication is conducted through the say command, and a privileged goalie client can also attempt to catch the ball. Two meta-level commands are also available: sense_body provides feedback on client status such as stamina and speed, and change_view selects a trade-off between quality of visual data and frequency of update. A detailed syntax for each of these control commands is given in Table 1 and Table 2.

• *Sensor information*: messages sent from the server to a client describing the current state of the game from the viewpoint of the client's player. There are two types of information, visual (see) and auditory (hear), as described in table 3.

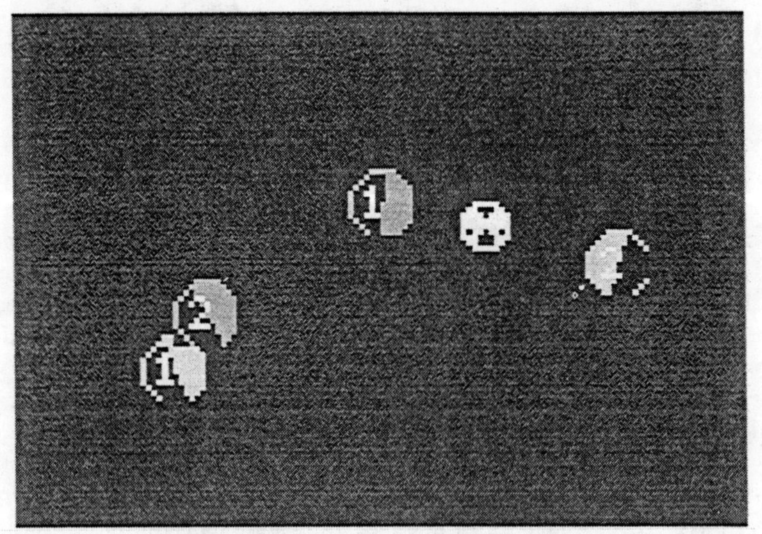

Figure 3. Close-up of players and a ball.

Soccer Server is a discrete simulation of continuous time. Thus, both the control commands and the sensor information are processed within a framework of "simulator steps". The length of the cycle between the processing steps for the control commands is 100ms, whereas the length of the step cycle for the sensor information is determined by the most recent change_view command issued by a client (we discuss the timing of these steps in more detail in Section 3). Note that all players have identical abilities (strength and accuracy of kicking, stamina, sensing) so that the entire difference in performance of teams derives from the effective use of the control commands and sensor information, and especially from the ability to produce collaborative behavior between multiple clients.

As a final feature, when invoked with the -coach option, the server provides an extra socket for a privileged client (called a coach client) that has the ability to direct all aspects of the game. The coach client can move all objects, direct the referee module to make decisions, and announce messages to all clients. This facility is extremely useful for tuning and debugging client programs, which usually involves repeated testing of the behaviors of the clients in many situations. An extension being considered for further versions of Soccer Server is a modified version of the -coach option that allows teams to include a twelfth client that has a global view of the game and can conduct sideline coaching during play by shouting strategic or tactical advice to players.

8.3.2. Referee Module: Playing by the Rules

The referee module monitors the game and, as in real soccer, regulates the flow of play by making a number of decisions based on the rules of the game. Firstly, when the ball is in a goal, the referee module announces the goal to all clients, stops play, and moves the ball back to the center mark. Secondly, when the ball goes out of the field of play, the referee module announces its decision (a throw-in, corner or goal-kick), stops play, moves the ball to an appropriate re-start position, and ensures (by physically moving them) that any opposing players are at least 9.15m from the ball. Thirdly, when the allowed game time for each half has been played, the server announces "time-up" and again stops play. Finally, once play has been re-started after a stoppage (by a successful kick command from the appropriate team), the referee announces that play has re-commenced.

At kick-offs, all players must be in their own half of the pitch. There is a period of 5 seconds for players to get back to their own half after a goal. During this time, the clients can use the move command to move their players directly to a specified position. After this, players in the opponent's half are moved to a random position in their own half by the referee.

In the first RoboCup tournament, the referee made no decisions about fouls. This was partly because the players had no hands (there was no goalie client) and partly because the offside rule had not yet been introduced. Remaining fouls, like "obstruction", are difficult to judge automatically as they concern players' intentions. The server therefore also includes an interface allowing a human user to instruct the referee to award a free kick. To do this, the user simply clicks on the desired restart position in the Soccer Server display window. The referee module then stops play and allows the user to choose between a free kick for either team or a simple "drop ball". Since such decisions about fair play actually call for significant understanding of the game, future versions of Soccer Server may include a dedicated socket for referee clients, thus encouraging research on automatic refereeing.

8.3.3. Message Board Module: Communication Protocol

The message board module is responsible for managing the communication between the players. A basic principle of Soccer Server is that each client should control just one player, and that communication between individual clients should only be carried out via the say and hear protocols introduced in table 1 and table 3. An important part of managing this communication is to construct an environment conducive to the efficient use of this protocol.

We saw in Table 1 that any player can broadcast a message to another player using the say command. In early versions of Soccer Server, this message was broadcast to *all* clients immediately as auditory information. Teams could therefore use this communication to sidestep the local nature of information in the game by, for example, having each player broadcast his own location and interpretation of the game at each time step. Another possible strategy was to use one client in a team (*e.g.*, a "captain") to direct and inform all the others. In the current Soccer Server, however, not all the clients are guaranteed to hear all the broadcast information, since there is a maximum range of communication of 50m, and any player can only hear one message from each team during each two simulation cycles. Also, the length of the message itself is restricted. Thus, effective and efficient use of the say command is encouraged, imparting information only when it is useful and timely.

Note that the server can connect with up to 22 clients, but that to facilitate testing it is possible for a single program to control multiple players (by establishing multiple socket connections). In a competitive situation such as RoboCup, such programs are only permitted on the understanding that the control of each player is separated logically. In practice, it is also not technically difficult to cheat the server and establish inter-client communication outside the server, so the "one-client one-player" rule is effectively a gentlemans' agreement.

8.3.4. Uncertainty

In order to reflect the nature of the real world, the server introduces various types of uncertainty into the simulation, as follows:

- *Noise added to the movement of objects.* The amount of this noise increases with the speed of the object.

- *Noise added to command parameters.* Small random numbers are added to the parameters of commands sent from clients.

- *Limited command execution.* The server executes only one command for each player in each simulation cycle. Each client program can get feedback on how many commands its player has executed via use of the sense_body command (*successful* execution must of course be monitored by the clients themselves; for instance, a dash has no effect if a player's path is blocked).

- *Inexact sensing.* The further an object, the less reliable the information returned about it from the server.

The presence of these uncertainties re-enforces the importance of robust behavior, and of the reactive monitoring of the outcomes of a player's actions.

8.4. Assessing the Soccer Server

How can we meaningfully assess the Soccer Server? In Section 1.2.3 we suggested that the main test of a good model is whether it provides answers to the questions we ask of it. Thus, since Soccer Server was constructed to enable the investigation of soccer, we could simply examine the quality of the conclusions learned about agent behavior in the domain. However, a more fundamental review is also possible. For instance, [Casti 1997a] summarizes several properties that can be used to assess models and simulations:

- *Fidelity*. The model's ability to represent reality to the degree necessary to answer the questions for which the model was constructed.

- *Simplicity*. The level of completeness of the model, in terms of how «small» the model is in things like number of variables, complexity of interconnections among sub-systems, and number of *ad hoc* hypotheses assumed.

- *Clarity*. The ease with which one can understand the model and the predictions/explanations it offers for real-world phenomena.

- *Bias-free*. The degree to which the model is free of prejudices of the modeler having nothing to do with the purported focus or purpose of the model.

- *Tractability*. The level of computing resources needed to obtain the predictions and/or explanations offered by the model.

Note that these qualities are more subtle than simply assessing whether a model faithfully captures all aspects of the phenomenon it represents. Let us examine the results of evaluating Soccer Server against each of the criteria.

8.4.1. Fidelity

One of the most important considerations during the development of Soccer Server was the level of abstraction for representing the client commands and the sensor information. One possibility was a low-level, physical description, for example allowing power values for drive motors to be specified. However, it was felt that such a representation would concentrate users' attention too much on the actual control of a players' actions, relegating true investigation of the multi-agent nature of team-play to the level of a secondary objective. Further, it is difficult to design a low-level description that is not implicitly based on a specific notion of robot hardware; for example, control of speed by drive motors is biased towards a physical implementation that uses wheels. On the other hand, a more abstract representation, maybe using tactical commands such as pass-ball-to and block-shoot-course, would produce a game in which the real-world nature of soccer becomes obscured, and in which the development of soccer techniques not yet realized by human players becomes problematic. Thus, our representation – using basic control commands such as turn, dash, and kick — is a compromise. To make good use of the available commands, clients will need to tackle both the problems of control in an incomplete information, dynamic environment and also the best way to combine the efforts of multiple players. Thus, we believe that Soccer Server achieves our goal of providing a simple test-bench with significant real-world properties.

The choice of abstraction level is also relevant to a further question that often occupied us during the design of Soccer Server: whether the simulator should be designed to model human soccer play or robot soccer play. This question has many facets, but in general the solution adopted in the Soccer Server is to be faithful to human soccer whenever possible. The justifications for this include the reasoning that real-world soccer is more immediately understood (and more attractive) to the casual observer, and that robot technology can change, whereas the nature of human soccer players is largely constant. Also, the closer the simulation to real-world soccer the more likely that knowledge acquisition from human expertise [Frank 1997] will be directly applicable.

Of course there are differences between real soccer and the model represented by the server, most notably the 2-dimensional nature of the simulation. Also, the simulation parameters are tuned to make the server as useful as possible for the evaluation of competing client systems, rather than to directly reflect reality (for example, the width of the goals is 14.64m, double the size of ordinary goals, because scoring is more difficult when the ball cannot be kicked in the air). This question of whether human soccer is being directly simulated is also important in the context of the overall RoboCup challenge, which we discuss further below.

8.4.2. Simplicity

The ultimate goal of RoboCup, as described in [Kitano 1997a], is "to develop a robot soccer team that can beat the Brazil world-cup team". Since this goal is well beyond current technologies, the RoboCup initiative defines a series of well-directed subgoals that are achievable in the short and mid-term. Implementing clients for Soccer Server is one of these challenges. The objective of RoboCup is that, as the level of sophistication of existing technology improves, the simplicity of the simulation should be altered appropriately.

In Soccer Server, one of the primary concerns about simplicity is the timing of the simulation cycles that govern the processing of commands and sensor information. The sensor information received by clients is rich, but the actions are simple. This is an argument for making the time between information updates longer than the time between action execution (*i.e.*, having an action processing cycle that is shorter than the information sensing cycle). If the information sensing cycle becomes too long, the ability to react to the opponents' actions becomes hampered. On the other hand, if the information sensing cycle becomes too short, the benefits of attempting to learn and predict the opponents' actions are diminished. The current lengths of these time steps are a compromise intended to strike a balance between these two conflicting goals.

8.4.3. Clarity

A large number of researchers have already used Soccer Server (we have already mentioned that the first RoboCup contest featured 29 teams from ten different countries). This is evidence that the workings of Soccer Server can be easily understood. In considering the "clarity" of the system, though, we also have to take into account the ease with which one can understand the lessons learned through the course of such research. So, we are interested in questions like, "What results have emerged out of the teams produced by the first RoboCup contests?" To some extent, the answer to this type of question is dependent on the efforts of the Soccer Server users themselves. However, the RoboCup initiative requires all the teams entering any tournament to write a paper describing their approach [Kitano 1998][Asada 1999]. These papers show that in the original contests, the programs that performed well were simple hard-coded systems with little learning ability. For example, at the Pre-RoboCup'96 tournament held in Japan in November 1996, the winning team was the Ogalets, which essentially relied on constraining each player to stay within a small, pre-determined area of the pitch and to choose between a small number of hard-coded passing directions when they could reach the ball. In the RoboCup'98 tournament held in July 1998, on the other hand, the winning teams were more sophisticated. The winners, from Carnegie Mellon University, used a layered learning approach, the runners-up used case-based reasoning and agent-oriented programming, and the third place team used reinforcement learning. Note that there is also a RoboCup Scientific Challenge Award, designed to encourage the application of novel or demanding techniques to the Soccer Server. At RoboCup'97 this was awarded to Sean Luke for his work on using evolutionary computation techniques to co-evolve a team of clients [Luke 1997].

As well as demonstrating the clarity of Soccer Server, it should be pointed out that sometimes the users of the system can show almost the opposite: that the system was not actually fully understood by the designers themselves. This happens when bugs are found in the simulation that can be exploited to the advantage of the client programs. One example of such a bug was an error in the implementation of stamina. Generally, the stamina of a player decreases when the player issues dash commands, thus limiting the player's ability to make further movements. However, it was found by some users that by using *negative* parameters in this command, a player's stamina could be increased!

A further flaw was found with the version of Soccer Server used for the RoboCup'97 contest, in which a problem with the handling of simulation cycles sometimes allowed players to kick the ball three times in very quick succession, thus imparting three times the normal "maximum" speed. However, it is in the spirit of the RoboCup contest that such bugs are reported to the designers, so that all users are working from an even footing. This spirit is also an illustration of the bias-free nature of Soccer Server.

8.4.4. Bias-free

Although developed solely at the Electrotechnical Laboratory, Soccer Server has benefited substantially from the opinions and suggestions of many users. In particular, there is a mailing list dedicated to discussing RoboCup, on which a consensus is usually reached before modifying the model represented by the simulation (this mailing list averaged over 200 messages per month between December 1997 and March 1998). To give a feel for the speed of change of the simulator, here is a partial list of some of the features changed since RoboCup'97: the introduction of the sense_body command, the ability to see the direction that other players are facing, the introduction of off-sides, the introduction of a goalie, and a re-modelling of the implementation of stamina. All these changes are designed to increase properties such as the similarity of the simulation to the human game, the importance of modelling the opponents' play, or the importance of modelling the teamwork being produced by a client's own team-mates.

Further evidence of the bias-free nature of Soccer Server is the presence of a public library of code (the RoboCup Simulation Code Archive, located at http://www.isi.edu/soar/galk/RoboCup/Libs/). Researchers can use this code to help reduce the development time of RoboCup clients. Some code is specific to particular situations, whereas some is general enough to provide infra-structure for complete agents (the code for the winning clients from RoboCup'97 is available). In particular, the archives contain the "libsclient" library, a collection of low-level functions intended to abstract away many details of tasks such as maintaining sockets, parsing the information from the server, and sending commands to the server.

8.4.5. Tractability

Soccer Server has to simulate a game of soccer and communicate with each client program in real-time. The system can run on a PC or a workstation with only modest memory and processor resources (although typically a user will also require at least one further computer to run the teams of clients). A more problematic limitation on Soccer Server in practice is the capacity of the network connecting the system to the clients. A large load on the network can lead to collisions that prevent client commands reaching Soccer Server or information being returned. Indeed, some teams at RoboCup'97 found it necessary to re-calibrate their software to the particular conditions found at the tournament site. However, it is difficult to predict the exact effects of changes in network conditions, so client programs with some flexibility in their interpretation of the state of play are encouraged. In general, this may not be an adverse property of the system; the need to understand adaptation and to cope with only local information were two of the prime qualities of complex systems identified in Section 7.2.1.

8.5. Focussing the Complex

We have described the basic Soccer Server, and shown that it rates favorably when compared against the criteria suggested by Casti. However, let us stop and take a step back. We saw in Figure 1 and figure 3 how Soccer Server renders its simulation of a soccer game with a full-pitch, overhead 2-D view. Since the height of the ball is not actually modeled in the simulation, this visualization has the undeniable benefit that none of the action will ever be missed. However, it is very different to the way that the majority of soccer fans experience the majority of their soccer: on television. The TV coverage of soccer typically incorporates a huge array of enhancements such as slow motion replays, multiple camera angles, mobile cameras on the touchlines, play-by-play commentaries, expert analysis, and on-screen statistics. These are all techniques that conduct what we will call *focussing* for the viewers. In contrast, the 2-dimensional display of Soccer Server makes no attempt to focus on any particular aspect of the game.

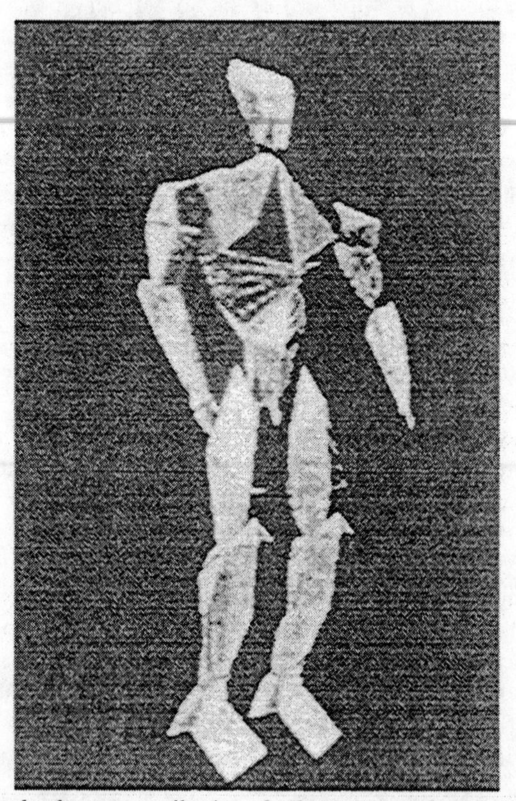

Figure 4. A single player as a collection of polygons (reproduced with permission).

This issue of focussing is related to clarity. What's interesting here, though, is that people watching a 2-dimensional Soccer Server game typically don't *notice* the lack of focus. That is – even when watching bad teams – spectators easily view the game as a single process, despite the large number of players and other variables involved.

So, focussing is not *necessary* to achieve clarity. But, the very fact that people seem to almost instinctively conduct such focussing when watching a Soccer Server game indicates that the issue is one worth investigating. We suggest that the basic architecture of Soccer Server, despite carrying out no focussing itself, can still be used as the basis for an investigation of focussing techniques.

One example of research that can be viewed as addressing the question of focussing is the work of Shinjoh and Yoshida on a 3-dimensional viewer system for Soccer Server [Shinjoh 1998]. This system takes the raw output of the Field Simulator module we introduced in figure 1 and represents it as an isometric 3-dimensional scene. Each player is realized by around 300 polygons combined together in an "imitation of the humans' muscle form" (cf. figure 4), and the actual interface is provided by a three-dimensional viewer called SPACE. A sample screenshot of SPACE is shown in figure 5.

Figure 5. A screenshot from SPACE (reproduced with permission).

The early versions of SPACE simply used five fixed cameras that could "pan" to capture the movement of the ball. There was also a simple set of rules for switching between these cameras. But, it was soon found that this basic control structure did not produce good results. For example, [Shinjoh 1998] reports that when watching the output of SPACE "most audiences... were bored", and that "because switching of the camera was done frequently, the audience was easily confused". More recent work on SPACE has concentrated on dealing with these problems by developing an Intelligent

Navigation System (INS) to control more carefully the selection of the displayed views. One of the principles of the INS control system is the incorporation of meta-rules such as "the reduction of audience confusion by the minimization of camera movements". The important feature of this rule is that it takes account not just of what happens in the virtual world but also assesses the information that spectators should receive, or may have received already.

Another area of research that requires the handling of focussing is automated soccer commentary. For example, the MIKE system [Tanaka-Ishii 1998] and the ROCCO system [Andre 1998] are both designed to monitor the output of the Soccer Server and produce real-time natural language descriptions of the game. For each of these systems, the commentary consists of a stream of (spoken or textual) utterances. Since only one utterance can be meaningfully stated at one time, a successful commentary system cannot avoid the issue of selecting which aspect of a game to focus on. Thus, new ideas on the focussing of complex systems will inevitably come out of such research. For example, one technique open to commentary systems is to analyze and comment on how the game is changing over time. Another common commentary technique is to make *predictions* of future events based on (past or present) observed behaviors or patterns of play.

In commentary, just as in the presentation of a 3-dimensional visualization, the appropriate information to convey next will depend on what the audience has been told already. In general, the flexibility to present a game to different audiences will therefore call for investigation of issues such as Grice's cooperative principles of conversation [Grice 1975] (*e.g.*, the maxim of quantity, that your contribution should be no more and no less informative than is required). Rules of communication become especially significant if a commentary *team* is being modeled, for example with an announcer following the ball-by-ball action and an expert providing higher-level analysis. There will also be an interaction between the different modalities of information (natural language and visual) when using such systems simultaneously.

This brings us to an important side-effect of research such as SPACE and MIKE: a more "life-like" realization of the virtual world represented by the Soccer Server. As we suggested above, the TV coverage of soccer has given most spectators fixed expectations about the coverage of a soccer game. These expectations tend to feedback into the development process when systems are evaluated by human observers. As an experiment, we created a video combining the 3-dimensional simulation of SPACE with the corresponding game commentary by MIKE. Even though these systems are both just advancing beyond the prototype stage, the result is a visualization that bears a surprising resemblance to the "realism" of actual TV broadcasts.

8.6. Conclusion: Towards a Theory of the Complex

We have discussed how the game of soccer is a good example of a complex system, well-suited to the task of modeling. We also described Soccer Server, clarifying the nature of the model it represents, and assessing it against the criteria suggested by Casti. Finally, we introduced the further criterion of *focus*, and described how researchers are using the basic architecture of Soccer Server to investigate appropriate focussing techniques for real-time complex systems.

Let us close by noting that the challenges represented by the domain of soccer have recently led to it being proposed as a new standard problem for AI research [Kitano 1997b]. Of course, the notion of a standard problem has long been a driving force for engineering research. Historically, for example, the acceptance of the "Turing Test" [Turing 1950] focused attention on the mimicking of human behavior as a test of machine intelligence. More recently, chess has received significant attention, generating important advances in the theory of search algorithms and search control, as well as motivating cognitive studies into the ways that humans approach the same problems [Levinson 1991].

Soccer (in particular contrast to chess) is a dynamic, real-time, multi-agent, system with incomplete information. We see Soccer Server as a virtual world that provides a tool for investigating such complex domains. It does this by providing a predictive model of the strengths of software algorithms for controlling agents in these environments. In the words of [Casti 1997b], Soccer Server is a "would-be-world" that "has the capacity of serving as a laboratory within which to test the hypotheses about the phenomena it represents". It is our hope that, for complex systems such as soccer, the efforts of researchers using the laboratory of Soccer Server will rectify the situation where "at present, there seems to be no known mathematical structures within which we can comfortably accommodate a description" [Casti 1997a].

References

Andre, E., Herzog, G., Rist, T., 1998, Generating multimedia presentations for RoboCup soccer games, in *RoboCup-97: Robot Soccer World Cup I*, edited by H. Kitano, Lecture Notes in Artificial Intelligence, Springer, **1395**, 200.

Asada, M. (ed.), 1999, RoboCup'98: *Robot Soccer World Cup II,* Lecture Notes in Artificial Intelligence, Springer (To appear).

Brooks, R.A., 1986, A robust layered control system for a mobile robot, *Journal of Robotics and Automation*, RA **2**.

Casti, J.L., 1997a, *Would-be Worlds: how simulation is changing the frontiers of science*, John Wiley and Sons.

Casti, J.L., 1997b, Would-be worlds: toward a theory of complex systems, *Artificial Life and Robotics*, **1**, 11.

Frank, I., 1997, Football in recent Times: What we can learn from the newspapers, in *Proceedings of the First International Workshop on RoboCup*, 75. Updated version appears in *RoboCup-97: Robot Soccer World Cup I*, edited by H. Kitano, Lecture Notes in Artificial Intelligence, Springer, **1395**, 216.

Grice, H.P, 1975, Logic and conversation, in *Syntax and Semantics: Speech Acts*, edited by P. Cole & J. L. Morgan, Academic Press, **3**, 41.

Huizinga, J., 1950, *Homo ludens: a study of the play element in culture*, Beacon Press.

Kitano, H. (ed.), 1998, *RoboCup-97: Robot Soccer World Cup I*, Lecture Notes in Artificial Intelligence, Springer, **1395**.

Kitano, H., Asada, M., Kuniyoshi, Y., Noda, I., Osawa, E., Matsubara,H., 1997a, RoboCup: A challenge problem for AI, *AI Magazine*, Spring, 73.

Kitano, H, Tambe, M., Stone, P., Veloso, M., Coradeschi, S., Osawa, E., Matsubara, H., Noda I., Asada, M., 1997b, The RoboCup synthetic agent challenge 97, *Proceedings of IJCAI-97* (Nagoya), 24.

Levinson, R., Hsu, F., Schaeffer, J., Marsland, T., Wilkins, D., 1991, Panel: The role of chess in artificial intelligence research, *Proceedings of the 12th IJCAI* (Sydney), 547.

Luke, S., Hohn, C., Farris, J., Jackson, G., Hendler,J., 1997, Co-evolving soccer softbot team coordination with genetic programming, *Proceedings of the First International Workshop on RoboCup*, 115. Updated version appears in *RoboCup-97: Robot Soccer World Cup I*, edited by H. Kitano, Lecture Notes in Artificial Intelligence,Springer, **1395**, 398.

Shinjoh, A. Yoshida, S., 1998, The intelligent three-dimensional viewer system for robocup, *Proceedings of the Second International Workshop on RoboCup*, 37.

Tanaka-Ishii, K., Noda, I., Frank, I., Nakashima, H., Hasida, K., Matsubara, H., 1998, MIKE: An automatic commentary system for soccer, *Proceedings of ICMAS-98*, 285. Also available from Electrotechnical Laboratory as Technical Report ETL-97-29.

Thompson, A., 1997, An evolved circuit, intrinsic in silicon, entwined with physics, Proceedings of ICES'96, *The First International Conference on Evolvable Systems: from biology to hardware*, edited by Tetsuya Higuchi, Masaya Iwata, and Weixin Liu, Springer-Verlag (Berlin), 385.

Turing, A.M., 1950, Computing machinery and intelligence, *Mind*, **59**, 433.

Table 1. Basic control commands.

(**move** *X Y*)
Move the player to the position (*X*,*Y*). The origin is the center mark, and the X-axis and Y-axis are toward the opponent's goal and the right touchline respectively. Thus, *X* is usually negative to locate a player in its own side of the field. This command is available only in the **before_kick_off** mode.
(**turn** *Moment*)
Change the direction of the player according to *Moment*. *Moment* should be -180 180. The actual change of direction is reduced when the player is moving quickly.
(**dash** *Power*)
Increase the velocity of the player in the direction it is facing by *Power*. *Power* should be -30 100. (The range is modifiable by the parameter file.)
(**kick** *Power Direction*)
Kick the ball by *Power* to the *Direction* if the ball is near enough (the distance to the ball is less than 1.0 + **ball_size** + **player_size**). *Power* should be -30 100, and *Direction* should be -180 180. (The ranges are modifiable by the parameter file.) The power actually achieved decreases from *Power* as the distance from the player to the ball increases and as *Direction* deviates from 0.
(**catch** *Direction*)
Try to catch the ball in *Direction*. The player (goalie) can catch the ball when the ball is in the rectangle which width is **goalie_catchable_area_w** (default=1), length is **goalie_catchable_area_l** (default=2) and the direction is *Direction*. This command is permitted only for goalie clients.
(**say** *Message*)
Broadcast *Message* to all players. *Message* is informed immediately to clients as sensor information in the (**hear** ...) format described in table 2. *Message* must be a string whose length is less than 512, and consist of alphanumeric characters and the symbols "+-*/_.()".

Table 2. Meta-level control commands.

(change_view *ANGLE_WIDTH QUALITY*)
Change angle of view cone and quality of visual information. *ANGLE_WIDTH* must be one of **wide** (=180 degrees), **normal** (= 90 degrees) and **narrow** (=45 degrees). *QUALITY* must be one of **high** and **low**. In the case of **high** quality, the server begins to send detailed information about positions of objects to the client. In the case of **low** quality, the server begins to send reduced information about positions (only directions) of objects to the client. Default values of angle and quality are **normal** and **high** respectively. On the other hand, the frequency of visual information sent by the server changes according to the angle and the quality: In the case that the angle is normal and quality is high, the information is sent every 150ms. (The interval is modifiable by specifying **send_step** in the parameter file.) When the angle changes to wide the frequency is halved, and when the angle changes to narrow the frequency is doubled. When the quality is low, the frequency is doubled. For example, when the angle is narrow or the quality is low, the information is sent every 75ms.
(sense_body)
Sense the status of body. The server returns the following information. **(sense_body** *TIME* **(view_mode** *QUALITY WIDTH*) **(stamina** *STAMINA EFFORT*) **(speed** *AMOUNT_OF_SPEED*) **(kick** *KICK_COUNT*) **(dash** *DASH_COUNT*) **(turn** *TURN_COUNT*) **(say** *SAY_COUNT*))

Table 3. Sensor information.

(**see** *Time ObjInfo ObjInfo ...*)
Transmit visual information. *Time* indicates the current time. *ObjInfo* is information about a visible object in the following format: (*ObjName Distance Direction DistChng DirChng FaceDir*) *ObjName* **::= (player** *Teamname UNum*) **| (ball)** **| (goal** *Side*) **| (line [l|r|t|b])** **| (flag [l|c|r] [t|b]** [*Num*]**)** **| (flag p [l|r] [t|c|b])** where *Distance* and *Direction* are the relative distance and direction of the object respectively, and *DistChng* and *DirChng* are respectively the radius- and angle-components of change of the relative position of the object. *DistChng* and *DirChng* are not exact values but are sketchy values. *FaceDir* is the facing direction of players. As the distance to another player becomes larger and larger, more and more information about the player is lost. Specifically, *UNum* is lost in the case of players further than a certain distance, and *Teamname* is lost from players who are very far. The quality of information is also changed by **change_view** commands. (See the description of **change_view** in table 2.) The frequency of sending this information depends on the quality of information transmitted. See the description of **change_view**.
(**hear** *Time Direction Message*)
Transform auditory information. *Direction* is the direction of the sender. In the case that the sender is the client itself, *Direction* is '**self**'. *Time* indicates the current time. This message is sent immediately a client sends a (**say** *Message*) command. Judgments of the referee are also broadcast in this form. In this case, *Direction* is '**referee**'. Possible judgments are: **before_kick_off, •kick_off_l, kick_off_r, • kick_in_l, kick_in_r, corner_kick_l, corner_kick_r, goal_kick_l, goal_kick_r, •free_kick_l, free_kick_r, offside_l, offside_r, •play_on,•half_time, time_up, extend, •foul**_*Side_UNum*, **• goal**_*Side_Point*.

Chapter 9
Feeping Creatures

Rodney Berry
Proximity Ltd.
rodney@proximity.com.au
http://www.cofa.unsw.edu.au/research/rodney

9.1. Introduction

"When we close our eyes, we open others."

Beyond the field of vision, beyond the frame of the screen, as far as the eye cannot see, stretches the realm of sound. Sound tells us much about the world around us. It finds its way through walls, around corners and across surprising distances. Sound is forever present in our lives because, unlike our eyes, our ears never close.

In this chapter, I will examine some of the possibilities for sound in virtual worlds. Although not an explicit "how-to" guide, there will be some treatment of techniques and technologies of virtual sound. My own recent work, *Feeping Creatures*, will be used as an example of some of the ideas presented. As a musician and sound-artist moving into the field of computer based virtual worlds, I hope that my personal perspective will provide some food for thought to other virtunauts.

9.2. Reality — Sensation, Perception and Cognition

Sound is so much a part of our whole sensory experience of the world, real or virtual, that it is meaningless to consider it in isolation. Because of this, I prefer to take a more synaesthetic approach to world making where all the senses are regarded as pieces of a puzzle. Together these pieces go to make up our view of reality.

A new-born baby appears to be such a synaesthetic being. Lights, noises, touches, smells and tastes are all an undifferentiated sea of sensation, lapping at the shores of our neonatal self. The baby responds accordingly with a general exuberant waving of limbs and primal burbling. As we grow older, we learn to sort and classify the various modes of perception, comparing one with the other to build up a model of our world. For the would-be builder of worlds, some attention to the processes of *sensation*, *perception* and *cognition* may well provide some clues about how to make these worlds more compelling.

Our senses alone do not give us a clear picture of the world. Our eyes blink and move all the time, the image on our retina shakes, pitches and rolls around like crazy, yet somehow, we perceive the room around us as stable. *Perception* is our way of organizing this overwhelming mass of sensory input into useable information. *Cognition* then takes this information, makes inferences, draws conclusions, builds models and generates meaning.

In this manner, our internal world model is assembled. Rather than re-build the whole model from scratch, moment to moment, we cheat. By retaining the old model and simply making small adjustments to accommodate new information, we experience the world as an ongoing thing. This is sometimes known as *conservation.* Conservation is just one of the many perceptual-cognitive principles that help smooth out the bumps and discontinuities of our perceived world. When we blink, the world is still there (isn't it?).

In pondering this, I wonder just to what extent we inhabit our models of the world in preference to the "real" one? Could it be that what we normally call 'reality' is not, in reality, reality? Is the real world, as we perceive it, just as "virtual" as the alternatives we are so busily constructing? If so, realistic and engaging virtual worlds may require more than bigger, faster, cheaper computers. They will almost certainly require a deeper understanding of how we come to organize our perception and why we accept the products of this perception as real.

For example, print on paper is now a very old technology which is supposed to become obsolete once the VR people get their act together. However, a well written book can cause us to imagine vivid images, smells and sounds. We can find ourselves lost in its world for hours at a time. Participants in text-based online worlds and games report a similar experience, to such a degree that their concerned families send them off for therapy, hoping that they will come back! Somehow, when the mind receives and processes these symbols, it connects them to previous experience and perceptions. Just as a computer may farm heavy tasks out to separate processors, a book co-opts the human brain to translate the symbols into a vivid experience for the reader.

In a virtual world, certain images, sounds and gestures may also have a symbolic role triggering some kind of extra response in the user/participant. When this happens, the user's mind is working to incorporate the virtual experience with its own inner symbolic landscape. When this ancient and powerful computer is called upon to help its silicon children, we suddenly move from mere interactivity to true engagement. Interactivity, pointing and clicking one's way through a database, is not enough to guarantee engagement. A microwave oven is interactive. Engagement comes through the perceived meanings of a virtual world, or how one interacts with the ideas behind and beyond its user interface.

But I am supposed to be talking about sound am I not? As I said earlier, we can never close our ears. Even as we sleep, we are processing a continuous barrage of auditory information. To cope with all this and make sense of it, we have evolved powerful methods of processing this sensory input in order to extract information vital to our survival in the environment. We can simultaneously listen to many sounds at once, choosing not to hear those which we deem unimportant. All day, we are unaware of the sound of the air conditioner until it suddenly turns off. At a party, we are able to ignore the ten conversations around us and just concentrate on one. We can judge the distance, size and direction of many sound sources, often without ever having to think about it. In computing terms, the processing involved is huge yet we can do this quite unconsciously.

9.3. Sound in Virtual Worlds

9.3.1. Real Sounds

Before losing ourselves in virtual sound, I would like to look at how sound manifests itself in the real world. Sound is essentially the passage of mechanical vibrations through a medium, for our purposes, usually air. Repeated movements of an object alternately compress and rarefy the surrounding air sending waves outwards like ripples on a pond. As the wave front spreads out, the sound *amplitude* (volume) decreases in proportion to the inverse square of the distance from its source. Put more simply, the sound level drops away a lot at first, then more gradually as you move further away from its point of origin.

It is probably obvious that we would use this change in level to judge how far away a sound is. We obtain cues about distance in a number of other ways also. Moisture in the air causes higher frequencies to become weaker with distance from the sound source making them sound more muffled when further away. Reverberation also provides distance cues. We hear, not only the direct sound, but the sound's many reflections off walls, the reverberant characteristics of the space. Not being part of the original sound, these reflections stay fairly constant. As we move further from a sound source, its reverberant reflections become more pronounced compared to the original sound. In addition, early reflections, the very first reflections to reach the ear, play a role in our subjective experience of distance.

The only truly quantifiable thing we can measure about sound is the relative density of the air in a particular spot at a particular moment in time. This is equivalent to the Amplitude of the sound. However, when we compare these densities at different moments over time, we can derive waveforms which exhibit the measurable qualities of frequency and wavelength. The shapes of these waves determine our subjective experience of such qualities as *timbre*(tone "colour"), *pitch*, *rhythm* etc.

Pitch is often defined as our subjective experience of frequency. Every doubling of frequency, we perceive the note to be the same, only higher, known as an octave. For example, *concert A*, the standard tuning note of most symphony orchestras, has a frequency of 440hz (cycles per second). The note *A* above it is 880hz and the *A* below it is 220hz. *Timbre* refers to our subjective experience of different wave-shapes as combinations of pure tones, or harmonics, of several different frequencies within the same sound. These combinations can change over time and allow us to identify different musical instrument sounds even if they are playing the same note as another.

9.3. 2. Digital Sound Manipulation

Sounds can be digitally sampled into a computer. Thousands of amplitude measurements allow us to reconstruct the waveform as a string of numbers (a CD-quality recording uses 44 100 such "samples" every second). The truly wonderful thing about this process though is that, once you have captured a sound as a series of numbers, that sound can theoretically be altered in any way at all, simply by altering the numbers. These changes can affect our perception of what is making the sound. This in turn affects what the sound may mean to us.

Pitch is one of the easiest things to manipulate digitally. The pitch or frequency of a sound gives us important clues about such things as the possible size of the sound source. For example, high pitched sounds are normally associated with smaller sound sources whereas low pitched sounds indicate a larger source of vibration (more massive things vibrate more slowly). Imagine the sound of our new-born baby again. If we are to alter the pitch and speed of the sound downwards, it seems like a much larger voice box is making the sound. In fact, the reasonably happy baby now sounds like a fully grown man having a serious nervous breakdown. The sound of a 1 meter long piece of wood thrown on the ground, if played back at half the frequency (one octave lower), will now sound like a 2 meters long piece of wood.

9.3.3. Spatial Perception of Sound

Knowing from what direction a sound comes can be a life-or-death survival skill. We achieve this sense of direction by comparing the different sensations received by each of our ears. Unlike light, sound travels at a leisurely pace of around 1000km/hour. This create small but discernible differences in the time a sound arrives at each ear from the one source. If the sound reaches our right ear before the left, we perceive the sound as coming from somewhere to our right. Stereo recordings exploit this effect either by using two microphones to record live sound, or by sending a signal in different proportions to the left and right channels in the mix. This is known as *panning*.

Discerning between sounds which come from in front or behind us is a lot more complicated. The outer shell of our ears is shaped to reflect sound back into the earhole. Our brain manages to interpret these subtle reflections off this shell to figure out whether sounds have come from behind or in front (the room's early reflections mentioned earlier can also provide valuable directional cues). When real sounds are recorded with a pair of microphones attached to a special dummy head, complete with ears, they can appear behind and in front of the listener who is wearing headphones. This kind of recording and playback is called "binaural".

9.3.4. Digital Spatialization

Filters based on laboratory measurements of sound perception can reproduce this effect on a computer. These are often referred to as Head Related Transfer Functions (HRTFs). First a person sits in the center of an acoustically dead space. A neutral burst of noise is played through a loudspeaker from a specific direction and picked up by tiny probe microphones inside the person's ears. By using a computer to "deconvolve" the neutral soundburst or "impulse sound" from the recording, we are left with the "impulse response" of the ears and head relative to the direction of the original sound. When a new sound is convolved with this impulse response, it will seem to come from the same direction as did the original impulse sound. By interpolating between impulse responses from points of a circle all around the person back in the lab, it is possible to make sounds appear to move around the headphone listener's head. the illusion works best with headphones because it is easier to control which sounds go in which ear. When listening through loudspeakers, the cross-talk between them combined with the re-filtering by the listeners ear cavities works to diminish the spatializing effect.

A number of ways exist to spatialize sound between a number of speakers instead of through headphones. Dolby Pro Logic has become a popular format in "home theater" type systems because it is used on many motion picture soundtracks. It uses a hardware decoder to extract four channels of sound from the incoming signal. These four channels include Left and Right and 'Surround' for the rear speakers. The fourth channel is a Center speaker which makes up for the 'hole in the middle' problem experienced with normal stereo where the listener is not seated at the center of the listening space (for example, most people in a movie theatre). Ambisonics is another popular method of spatialization although it is not common among domestic systems. Ambisonics use four channels of encoded audio which can be decoded into a variable number of speakers placed around the listener. With arrays of eight, sixteen or more speakers, the sound can appear above or below as well as anywhere around the listener. The listener needs to be located at the center of the space.

The simplest spatialization method of all is to simply use four channels of sound panned between four loudspeakers, two at the front and two at the back. This requires the listener to be at the center of the room and only works on the horizontal plane (unless you add up and down channels as well!).

Systems exist which can process sound so its spatial information can be extracted by a Pro Logic decoder, fed into an ambisonic speaker array or for binaural headphone listening. For builders of virtual worlds, this can open up huge possibilities. At the High end of the market there is the Lake DSP Huron workstation which can spatialise several channels of live real-time audio. A reasonably fast PC or workstation computer can offer limited sound spatialization depending on its hardware resources. Binaural spatialization is implemented in the VRML2 modeling language [VRML2][Spatialization].

9.3.5. Sound in VRML2

In VRML2 the Virtual Reality Modeling language, there are already some limited but effective provisions for dealing with sound and its behaviors in a virtual space. It is important to remember that VRML does not attempt to accurately model the intricacies of sound in the real world. It merely provides tools to allow us to approximate it with the aid of a little bit of ingenuity. Because the language is still under development, we can expect much more versatile features in the future.

Books and Web tutorials on VRML2 abound [Brown 1997] so I will only deal briefly with how it incorporates sound. A VRML world is made up of various specialized "nodes". The parts of the VRML2 language that deal specifically with sound are the Audio Clip node, the Sound node and the Movie texture node.

The audio Clip node is used to retrieve and play a sound file from a specific URL. It lets you determine when the sound file starts playing, how long it lasts and whether or not the sound should be looped. The start and stop times can be controlled externally by sending outside events to the Audio Clip node. The pitch or speed of playback can also be adjusted here. The Audio Clip node is able to keep other nodes informed of the duration of playback and whether or not it is in operation at any given moment.

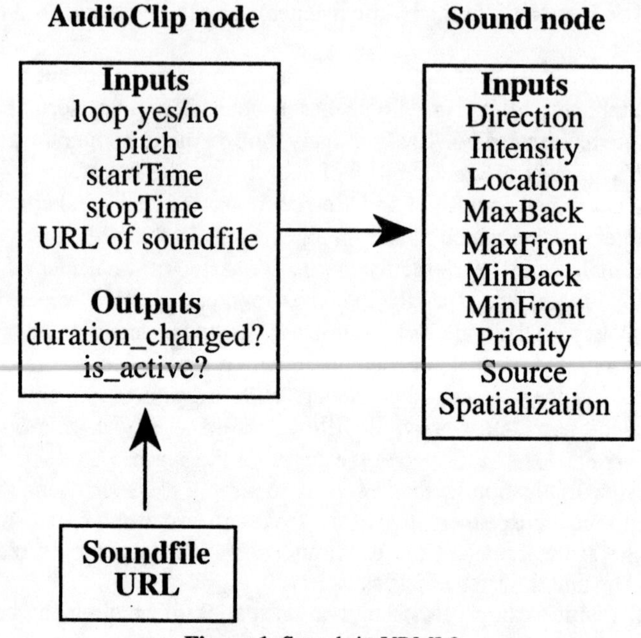

Figure 1. Sounds in VRML2.

So far though, we have what amounts to a rather unwieldy sound file player. The really interesting stuff happens within the Sound node. Using the Audio Clip node as a source, the Sound node handles the spatial aspects of sound. Imagine we place a loudspeaker in the world and want it to emit sound. In the real world, the sound will be much louder for a greater distance in front of the speaker than behind it. The sound node simulates this in the following manner. Firstly we specify the Location of the sound source in the world, then we use the Direction field to indicate which way it is facing. Several other fields let us determine the extent and the shape of the sound itself. To define how far in front of the speaker we want the sound to be heard, we set the *maxFront* field. To define how far behind the speaker the sound extends, we set the *maxBack* field.

These two fields mark the bounds of an ellipsoid shape surrounding the point identified as the source's Location. If both fields share the same value, the sound's area of influence will take a spherical shape. The remaining two range parameters, *minFront* and *minBack*, define an inner ellipsoid within which the sound plays at maximum intensity. As we approach the speaker from the front, no sound is heard until the boundary defined by *maxFront* is crossed. the sound gradually gets louder until we reach *minFront.* As we move on through the speaker, the sound remains at maximum intensity until we reach *minBack* from where it rapidly fades out until *min-Back* is reached and no sound is heard.

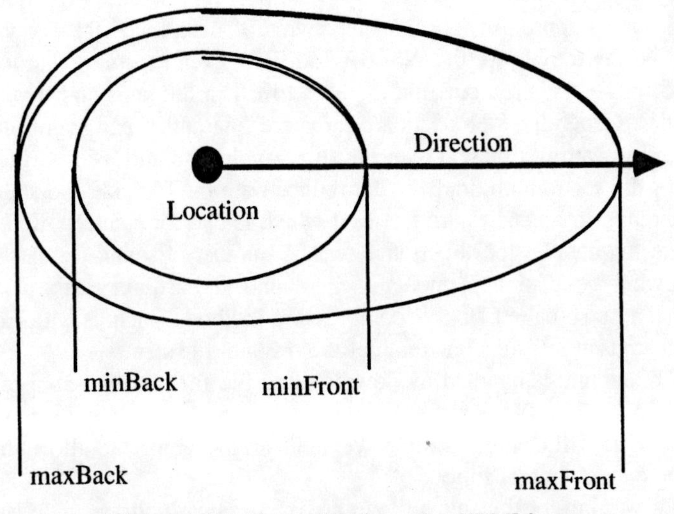

Figure 2. Sound spatialization using VRML2.

VRML2 departs from real-world acoustics when we try to put such things as walls within the ellipsoids. There is no way of telling the browser that the sound can not pass through the walls. Instead we have to work around the problem by using a proximity sensor to turn off the sound as we enter the prescribed area. Of course you were about to ask "...but what about my river of molten lava? It's sound is not an ellipsoid, it is long and narrow". Once again, we have to be inventive and use several overlapping Sound nodes to approximate other shapes. Here the Sound nodes would share a common sound file to correctly locate the molten lava sound. Another problem is the effect of distance on the sound. In the real world, as I mentioned earlier, high frequencies become less audible as we move further from a sound source. We can approximate this effect by gradually cross-fading the sound with another, slightly muffled version of itself as the user moves away from the sound source.

When the *Spatialize* field is active, the VRML browser is instructed to place the sound appropriately in the stereo mix between the left and right speakers. Provision is made in VRML2 for the use of HRTFs to give a 3D sound image when headphones are worn. In practice, the number of sounds which can be simultaneously spatialized in this way depends on the speed and memory resources of the host computer. The quality of the computer's sound hardware also places limits on the amount of spatialization available. HRTF-based spatialization places greater demands on computing power than simple stereo panning. for this reason many browsers let you switch between headphone speaker-oriented listening. Sounds which have no particular location in the space, background music for example, are played with spatialization disabled.

9.3.6. Size of Sound Files

When working with digital audio, it doesn't take long to realize that sound files can be obscenely large and take forever to download over a network. A CD quality stereo recording takes up approximately ten megabytes of disk space for every minute of sound. That is way too big for the World Wide Web, even assuming that it gets much faster in the future. Luckily, our mind is able to fill in the gaps and to smooth over discontinuities. Generally, web users are far more tolerant of bad sound quality than they are of long waiting times. Lowering the sample rate and using 8 instead of 16 bits to record the sound both dramatically reduce both the file size and sound quality. As a rule, the high frequencies are lost first which is bad for spatialization as we rely on these high frequencies for directional cues. A number of sound formats reduce file size by throwing away some elements of the sound that are not easily discernible to the listener. These so called "lossy" compression methods, such as MPEG, can yield reasonable sounds at less than one tenth of the original file size.

In VRML2, sound is handled by downloading the files located at specific URLS then playing them. The file is not heard until it is completely downloaded and saved to your disk. This will change as future versions are developed to allow streaming of both audio and video in real-time.

Instead of waiting for the entire file to arrive, streaming allows us to begin listening to the first part of a sound or video file while the rest of the file is still being transferred. This creates the illusion of speed and even allows live sound to be broadcast over the web. This means live real-time voice communication between participants in multi-user worlds. It is reasonable to expect that multiple audio streams will eventually be spatialized by future versions of VRML. Also, other kinds of information will be able to be streamed including geometry data and MIDI.

9.3.7. MIDI

MIDI (Musical Instrument Digital Interface) is another way of sending sound information down a wire. It has been in use for nearly twenty years and has become an established standard for music synthesizers and samplers. In much the same way that VRML describes geometry instead of individual pixels, MIDI consists of instructions for playing music rather than sending the sounds themselves. It can be sent as a continuous stream or as a file containing MIDI data. MIDI files can be relatively small so they can be loaded quickly. The disadvantage, of course, is that we are dependent on the quality of the machine reading the file, or receiving the stream, to re-construct the sounds as originally intended.

MIDI was originally developed for the networking and remote playing of electronic musical instruments. MIDI instructions tell a synthesizer or sampler such things as when to start playing a note, what note to play, how hard the key is hit, what sound to use, when to stop playing the note etc. By linking a computer to an external synthesizer, a very high quality of sound is possible without the computer having to do any actual sound processing. There is provision in MIDI to send program changes to select which instrument sound will play. The problem here is that MIDI instruments each have a very different set of preset sounds as well as new sounds which the owner has created. To overcome this problem, a standard was agreed upon called General MIDI.

Many newer synthesizers and computer sound cards incorporate a set of General MIDI sounds. This is simply an agreed-upon matching of program numbers to specific kinds of sounds. For example MIDI program 1 is a Grand Piano in the GM set and will be true of any GM instrument no matter who made it. However, within this standard there is some variation and a MIDI file will sound slightly different between synthesizers, especially for sounds like "poly-synth" whose meaning is open to interpretation.

General MIDI is mostly used for background music with VRML and other web-based applications as it is difficult to do anything really interesting with the sounds available. At this stage, spatialising and streaming MIDI are not viable in VRML2. It is possible to play only one MIDI file at a time and only from the beginning of the file. Future versions of the VRML language will no doubt have expanded MIDI capabilities.

Leaving the world of VRML and the Web for the time being, MIDI still has a lot to offer. If the world we are making is to be viewed from just one computer with a single user, MIDI can dramatically improve our sound options.

Although newer sound cards can often include high quality MIDI synthesizers, a separate, dedicated MIDI synthesizer or sampler can offer many extra avenues for exploration. Such devices can store a number of sounds in RAM and ROM ready to be activated by MIDI commands. Extremely fast and fine control of the sound is possible in real-time and a wide range of outputs and other options are available. For those developing their own software, many development packages include libraries for dealing with MIDI and most serial interfaces can be persuaded to send and receive MIDI data.

In addition to Note-on, Note-number and Note-off etc., there is a large array of separate continuous controller messages which can be used to control various parameters within a MIDI device. Some are always associated with particular parameters, for instance, controller number 7 is usually linked to MIDI volume and controller 10 is usually used to control panning between left and right. Other controllers, such as number 12 and 13 are left unassigned and can be used to control anything that is able to receive these instructions.

In a virtual world, a looped sample of a telephone ringing might be activated by sending a MIDI note-on command to a sampling synthesizer. A program change message or note number could be used to select the correct sound. As we move toward the telephone in our virtual world, MIDI volume commands make it sound louder. If we run quickly past it, MIDI pitch-bend could be used to simulate the doppler effect by adjusting the pitch according to our speed of motion. As we move further away from the telephone, not only is the volume turned back down but a filter, controlled by an especially assigned controller, could gradually make the higher frequencies more muffled just like in the real world. In addition, another controller could be used to adjust the balance between the straight telephone sound and the same sound fed through a reverb effect.

Those not especially concerned with realism or reality could of course map drastic warping and distortions of the sound to events and variables within their virtual world. Gods will be gods.

A sampling "workstation" style instrument can easily hold hundreds of short sounds ready for activation by MIDI. MIDI uses 16 distinct data "channels", each capable of controlling an independent instrument. Modern MIDI devices can usually receive and play from many channels at once, each with its own separate set of continuous controls, program changes etc. Groups of sounds are often assigned to one MIDI "program" on the one channel. These sounds are mapped out across the keyboard range so that different note numbers activate different sounds. Often referred to as a "multi-sample" or a "key-map", this method is good for situations where you want to trigger a wide range of sounds quickly. This is the case for drum kits and multiple samples throughout the range of an instrument such as the piano.

Controls, such as pitch-bend, volume and panning, affect all of these key-mapped sounds uniformly. Two other parameters offer more independent control of key mapped sounds because they are local to single note events. One is Velocity, which simulates how hard we hit the keys. The other is Poly-touch or polyphonic aftertouch which reacts to how hard a key is pressed and held after it is struck. Both can be mapped to any assignable parameter within a MIDI instrument.

MIDI program change messages assign different sounds or key-mapped groups of sounds to the channel on which they are sent. Although synthesizers are quite fast to respond, sometimes there will be glitches if sounds are still playing when program change messages are received.

To sum up, MIDI offers a fairly stable and standardized protocol for controlling sound. Using dedicated external MIDI devices, MIDI can be a powerful ally for dramatic real-time sound. As computers become more powerful, software-based MIDI instruments will enable a great deal more flexibility and portability than their expensive hardware cousins. Over the net, MIDI is a fast way to deliver fairly un-exciting background music with very small file sizes. As VRML and other methods of delivering networked virtual worlds expand their standards, MIDI will become much more useful and versatile.

MIDI is now about twenty years old and there is talk of working it into a whole new standard. It has been suggested that, instead of the old serial cable connection, a fast 'fire-wire' based interface will be used. This would enable the transmission of live streaming audio and many other goodies. The major challenge will be to make the new standard somehow compatible with all the various existing MIDI devices and systems.

9.4. Feeping Creatures

MIDI has been used to great effect in my own virtual world called *Feeping Creatures*. The name *Feeping Creatures* is a spoonerism of *Creeping Features*. The term refers to a tendency among software developers to overfill a new program with features until it becomes a bloated unworkable mess. The title was chosen as a reminder for myself and the programming team to keep the piece as simple as possible. The ideal was to see how much variety and complexity could be generated with the absolute minimum of complication.

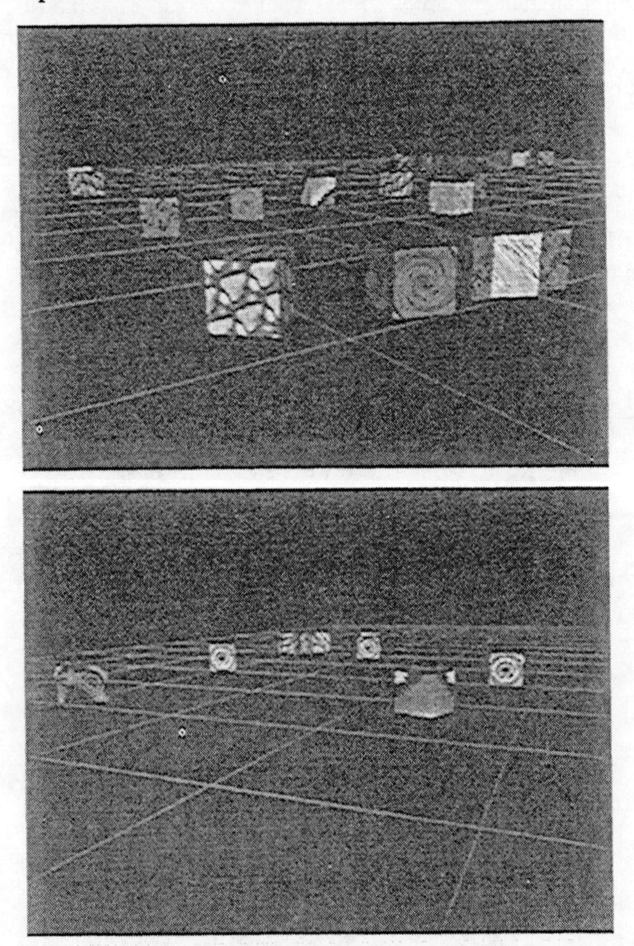

Figure 3. Two views of Feeping Creatures.

I make a distinction here between complexity and complication. *Complexity* is what emerges in nature as a result of interactions between essentially simple interdependent elements. *Complication* is what comes from human attempts to create the outer appearance of complexity without addressing the underlying process. The result might look similar but the point is lost. One approach takes a block of wood and tries to carve a tree out of it, the other simply plants a seed and waits.

9.4.1. Feep's World

The *feeps'* world is a flat green grid across which the *feeps* and the observer move (cf. figure 3). Their shapes are simple cubes covered with moving textures. Food is represented by green triangular pyramids (*trees*) which grow up from the floor of the grid. The user of the program moves a mouse to steer a virtual camera with its attached virtual microphone across the grid. The screen shows the view from that camera while the speakers play the sounds collected by the microphone.

9.4.2. Feeps

Each *Feep* has a sequence of musical pitches which form its chromosome. These are mapped to MIDI note numbers which are sent to an attached synthesizer or to the computer's built-in sound generators. When two *Feeps* mate, portions of each parent's note list are passed on to the offspring to form a new chromosome or pitch series.

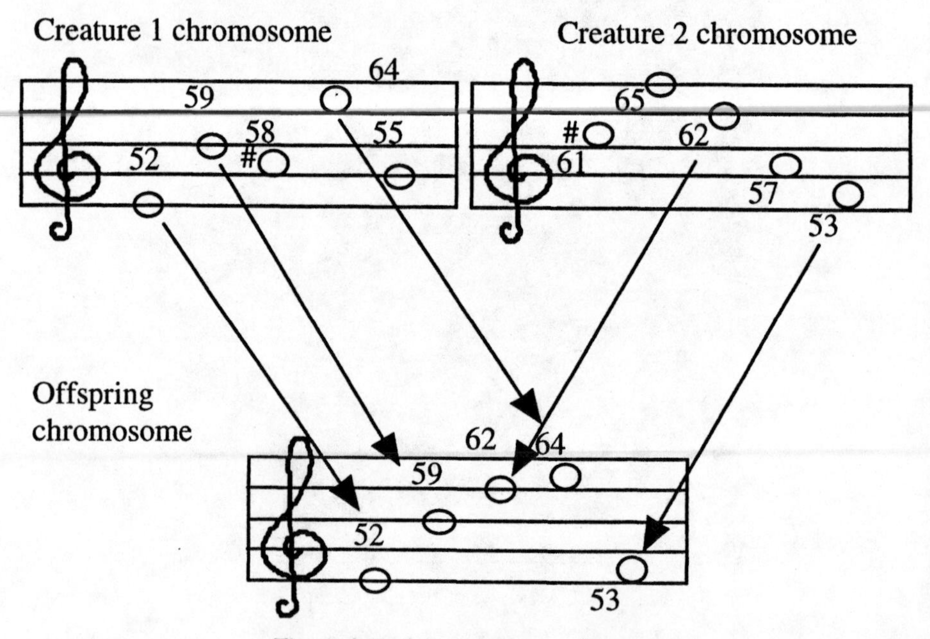

Figure 4. Evolving genotypes of Feeps.

At birth, a *feep* is randomly assigned a numerical value determining its preference for mating. If this value is high, the new *feep* will seek out partners which are, on average, musically consonant to its own note series. If the value is low, its preference is for those more dissonant to itself. In musical terms, we are only dealing with average vertical relationships between two lists of notes. This value is found by first finding the difference between the first MIDI note numbers of each series, then dividing the result by 12 and keeping the remainder (modulo 12). This returns a value of less than 12 which is then compared with a hierarchical table of intervals. The current table gives the unison or octave a value of 0 (most consonant) and the semitone a value of 11 (most dissonant). The remaining intervals fall between these extremes in an arbitrary order. The process is then repeated between every note in each list before averaging out the results. By tabling the results of these calculations, the program can keep track of who will mate with whom.

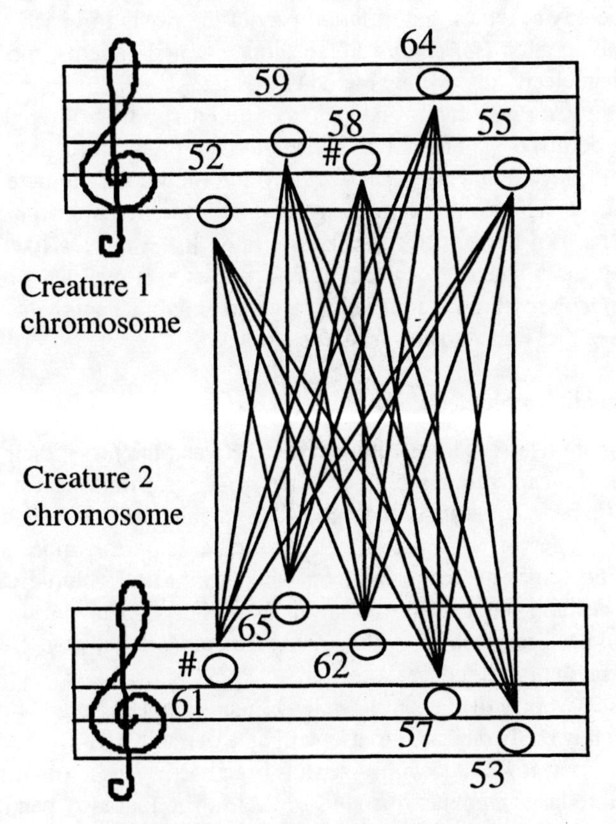

Figure 5. Harmony between potential mates.

This method provides a fast way of getting a general sense of harmony between potential mates[1]. This method would be easy to adapt to microtonal even-tempered scales by substituting a new hierarchy of intervals and modulo value. The value would be based on 19, 64 or whatever instead of the equal-tempered twelve note scale normally employed in western music.

I have yet to find a similar method for distinguishing between rhythmic material. I am considering setting up some way of distinguishing whether rhythmic groupings divide evenly into two, three, four, etc, to make general judgements about their over-

[1] Although not implemented at this time, it would be possible to find rough horizontal relationships between the *feeps'* pitch series using a similar method.

all feel. The problem with making all these calculations in real-time is that each algorithm must be very economical so as not to bog down the computer with trivia. Fortunately, individual precision is not necessary here. As long as general trends can be discerned across a population, quick and dirty solutions are the most suitable.

Rhythm is dealt with separately to pitch. This allows repeated cells of melodic material to cycle with rhythmic figures of a different length. This causes different parts of the melody to return in a different part of the rhythm in each subsequent cycle. These note/duration *isorhythms* create short repeated patterns, more reminiscent of bird calls than clearly discernible melodies.

In *Feeping Creatures*, rhythm is analogous to energy flow through the ecology. Each item of food (*tree*) contains one *duration* value (how much time passes before the next note is played) and one *articulation* value (how long the note sustains once initiated). When a *feep* finds a *tree* and eats it, it adds the *tree's duration* and *articulation* value to its own rhythm list. Unlike the pitch list, which is fixed at birth, the rhythm list increases when a *feep* eats and decreases as a *feep* ages or fails to find food. When a *feep's* rhythm list falls below a given length, meaning its energy level is too low, the *feep* dies and vanishes from the world.

9.4.3. Kinaesthetic Implementation

The main program oversees all the interactions between the *feeps*, their positions and the position of the camera/microphone. When a *feep* is within hearing range of the microphone, the program assigns it a new MIDI channel and timbre according to its *species*. Each species of *feep* also has a distinctive texture mapped onto its cubic shape. As a *feep* comes nearer to the microphone, its MIDI volume value increases causing it to become louder. The program also adjusts sound levels between left, right, front and back for each individual *feep*. This gives every *feep* a distinct spatial location between the four loudspeakers.

Although the work will run on an SGI O2 using the computer's internal synthesizer software, it is really designed to use with an external MIDI synthesizer to handle the sound. A Kurzweil K2000 synthesizer has been used because of its flexibility and its ability to store large amounts of sound in RAM. The four-way panning of sounds is achieved by building each "instrument" on the synth in two identical layers. One layer is assigned to the front left and right output channels and the other to the back. MIDI controller 10 pans both layers between left and right while controller 12 is used to crossfade between the front and back layers. doubling the layers in this manner allows only 12 of the kurzweil's 24 voices (available simultaneous notes) to be used. However, each of these 12 voices can be independently spatialized and assigned separate instrument timbres.

The four-way sound, coupled with the projected video screen and basic mouse interface, helps to create a sense of being drawn in to the imaginary world of the *feeps*. A great deal of care was put into the way sound, visual and kinaesthetic information go together in the work. For instance, creatures visible on the screen in front of the user are heard to be in that position also. A creature behind the user will be heard from the rear speakers and, if moving to the front, will enter the field of view as its sound moves forward. This is analogous to the way we experience sound in the real world. Its internal logic creates a feeling of being somewhere.

Much of our kinaesthetic sense involves a convergence of cues from our senses of hearing, sight and touch. This *cross-modal cuing* between different modes of perception is probably the most crucial factor influencing one's degree of immersion in a virtual environment. In *Feeping Creatures,* the mouse provides the kinaesthetic link between the user and the work. When the mouse is moved, the viewer's position and orientation are immediately altered. The program provides instant feedback to the user in the form of changing sound and images. This enables the user to build his or her own internal model of the world which extends far beyond the plane of the screen.

9.4.4. Physics of the Virtual World

In a virtual world, everyday physics can be ignored. It is possible to move instantly from one place to another, to make capricious changes to the fabric of reality. After all, such things are nothing but numbers and are ours to command. However, even our most spectacular dreams and delusions have their own internal logic. Humans have a desire for their inner and outer worlds to be congruent. Engagement demands continuity and consistency. The suspension of disbelief requires at least a seed of reality. To this end, the *feeps'* world comes complete with its own laws of physics and biology.

The physics of *Feeping Creatures* affect both the *feeps* and the user alike. All objects in the world, including the user, move as if affected by friction and momentum. *Feeps* always align themselves along their line of travel. They accelerate, decelerate and turn as if they have real mass. They head in the direction of food or toward a prospective mate according to when these needs were last met. This apparent purposiveness also helps to deepen one's sense of engagement with the work.

The actual method of moving through the *feeps'* world is simple. Pushing the mouse forward increases forward motion, pulling it back slows the camera to a stop before starting to move backwards. Moving the mouse left or right changes direction of movement accordingly. Although the user can move freely around the environment, he or she cannot actually make changes to that environment, nor do its inhabitants show any awareness of the observer's presence. The participant's role is that of an *invisible tourist*, a mute observer of evolving sonic cultures.

Feeping Creatures' database is in a state of continual flux, objects come into being, grow, move around and are gone. The work, while keeping its general character, is never quite the same in successive runs of the program, nor does a user ever reach a point of having seen all of the work. The user leaves the work when he or she has had enough. In this way, the work creates a sense that its world continues to exist, even when the user is no longer present. In seeking to build more autonomy into my artworks, I find myself working towards an idealized *perpetual novelty machine*. Left to its own devices, a perpetual novelty machine would forever continue to produce new and unique output. In nature, the process of evolution exhibits this creative quality. Nature seems capable of an endless variety of forms and behaviours without outside intervention. For this reason I have been fascinated by the potential of evolutionary programming and Artificial Life to bring these life-like qualities into human artefacts.

9.4.5. Rogue Tree Phenomenon and Hillbilly Effect

The work was first shown at The Performance Space gallery in Redfern, Sydney as part of the Australian Perspecta exhibition in August 1997. Having an opportunity to observe it for a long period of time was hugely rewarding.

Different music came out of people's different ways of moving through the world. Some systematically followed one creature, causing a central theme to underpin the other changing lines of music. Others (myself included, I confess) preferred to charge full-speed through large groups of creatures to hear the resulting dense cloud of *feep-songs* wash from one end of the room to the other.

Observing the work itself, certain recurrent patterns became apparent. One of these was the *rogue tree* phenomenon. On rare occasions, one of the normally stationary *tree* objects would start to roam the landscape devouring others of its kind, playing its own homicidal little tune as it went. Although an easily corrected software glitch, the rogue tree became so popular that we decided to leave it in the program as a "feature".

Another development in successive runs of *Feeping Creatures* was the *hillbilly effect*. *Feeps* are born with a randomly assigned preference for others more consonant or dissonant to themselves. For this reason, genetically similar individuals will eventually come together who also have a preference for a very high level of consonance. These will be most attracted to members of their own immediate family. This increases the chance of their offspring doing the same, leading to a sudden explosion of genetically identical *feeps*. In the exhibition, these *hillbilly* groups quickly outbred the other more divergent types and totally monopolized the food supply. Soon nothing remained but a large population of identical *feeps* all playing the same tune. The effect did not manifest itself for eight or more hours but, as soon as it chanced to begin, the world quickly ran down into a state of changeless repetition.

In the initial design, attempts were made to avoid in-breeding by not allowing a *feep* to mate until it reached a certain age. It was hoped that by then the new creature would be far enough away from its relatives to find less similar partners. This precaution helps delay the onset of inbreeding only partially[2].

9.4.6. Wish List for a New Version

The development of *Feeping Creatures* will continue for another year or so. In the near future, the following will be explored and incorporated:

- Predators (*feep-lions?*) will be introduced to prey upon the *feeps* and take on rhythmic characteristics of their victims.

- The environmental model will be expanded to include further aspects of natural systems. There really needs to be a few more feedback loops where one aspect of the system affects another which in turn affects yet another. I would like to in-

[2] One of the programmers hit on the idea of making a *feep*'s voice break by dropping its pitch two octaves when it reached *puberty*. That is why the small creatures tend to have a higher pitched song than the older ones.

clude more ways for the creatures to listen to one another. For example I would like the predators to hunt by analysing rhythms.

• The environment will become more complex, possibly including weather and terrain. This will force *feeps* to evolve and develop more elaborate survival strategies in response. The *hillbilly effect* will be eliminated by permitting larger populations with more complex patterns of breeding and interaction. In addition, I intend to work more closely with theoretical biologists and ecologists to find ways of generating more bio-diversity in a small environment.

• The world will no longer be flat. Flying *feeps* may appear.

• The user will no longer be an invisible tourist in the *feeps'* world. Creatures will become curious and flock around the camera, or run away if startled by a sudden movement from the user. The user may even be able to practice a form of farming by capturing and breeding selected *feeps.* in a separate enclosure.

• The work will be made compatible with the Lake DSP Huron audio spatialization workstation and similar technologies. This will enable a more realistic spatial sound image of each *feep's* position and movement around the listener.

Also, some more adventurous (and expensive) future possibilities include the following:

• As domestic computers and game machines improve in power, it will be possible to adapt the software to run on such a system. This would open up the prospect of a much wider audience and even the possibility of selling it as an entertainment product.

• A live musician could play into a microphone and have the computer spawn new creatures based on the notes played. As these new *feeps* reproduce, the player could play along with what he or she hears from the speakers, further changing the environment.

• A network of computers and monitors could be placed throughout a large ensemble or symphony orchestra. Scrolling manuscript notation on the screen would direct the players to play a live version of the *feeps'* songs.

9.5. Some Final Words

I think that sound will improve along with the technology for the visual, tactile and kinaesthetic aspects of virtual worlds. Virtual acoustics and physical modeling of sound will become more commonplace as computing power increases. Instead of complicated tricks with the volume control and position sensors, future worlds will require only descriptions of the dimensions and materials of virtual rooms to deliver a faithful simulation of their acoustics. Physical modeling of events such as collisions of objects will automatically result in sounds without the need for any sound files to be loaded.

Whatever wonders these exploding technologies deliver up, the bottom line will still be how much the imagination of the world builder stimulates and inspires the imagination of the world user. The capacity for sound to carry meaning is not necessarily increased by technology unless ways are found to connect the products of this technology with our own inner symbolic landscapes.

References

Brown, G., 1997, Late Night VRML2.0 with Java, edited by B. Roehl, J. Couch, C. Reed-Ballreich, T. Rohaly and G. Brown, Ziff-Davis press (Emeryville), 188.
VRML2, http://www.dform.com/inquiry/tutorials/vrmlaudio.
Spatialization, http://www.dform.com/inquiry/spataudio.html.

Chapter 10

Art in Virtual Worlds: Cyberart

Olga Kisseleva
Artist
kisselev@cnam.fr

10.1. Introduction

The present world is characterized by the explosion of new technologies of communication and by their unexpected uses, that contributes to modify our behaviors and our way of thinking. Facing this evolution, we attend the emergence of a new form of art that positions itself in relation to these technologies, not merely by the tools that it uses but through the questioning that it causes.

The Contemporary Art is badly perceived by the public, who gets lost in the different states of artistic activity and is, however, incited to consider it as an essential element to his integration in the current society.

In the 20th century, the Modern Art has given up its place to the Contemporary Art. Its role and its place in the society have changed. While the artistic century opened up with the experiments of the limits (Kazimir Malevitch, Marcel Duchamp) and broadened its territories and problematic endlessly; the invention of the photography and then of cinematography have modified deeply our relationship with the reality, the time, the memories, the facts and the possible.

In the 20th century, art had first questioned its proper constituents, then its domain of validity, and finally its fields of investigation. Since then, all the categories are explored and studied. The social sciences, the philosophy, like the exact sciences, are questioned. The narration, the spectacular, the visible and intangible belong, henceforth, to the domains of creation. But also, the daily, the other, the world do.

We can identify our society's state of change and consider that we are living an era similar to the one marking the end of Middle Ages and the beginning of the Renaissance. The sciences and the art of the 20th century are elaborating a new and richer paradigm, explaining better the Universe. This rupture can be assimilated to a new Renaissance. And fundamental changes will undoubtedly emerge from this collective exploration.

10.2. Cyberart

The traditional artistic activity (for instance Modern Art) consists of a representation of the reality, in a manipulation of materials, intended to create tangible mirrors of our experiences and desires.

Contemporary Art does not reflect life. It prefigures life. By now, thanks to the mechanisms of new digital and communication technologies, the masterpiece itself can become a manipulation of reality : a digital and non-material space in which we can literally penetrate. If Modern Art belongs to the consumer system, Contemporary Art belongs to the communication system.

Contemporary Art precedes life. According to Marshall McLuhan, the artist is the effect preceding the cause.

Contemporary Art explores the change of paradigm, with a perception parallel to the scientists' one. We attended to the break between the classical perspective and the impressionist movement, to the appearance of movement and color in the futurist' work; to the appearance of relativity in the work of Marcel Duchamp; to the creation of new spaces by the cubists.

The introduction of the space-time notion, in art, is undoubtedly, one of the most revolutionary events, ever met by our culture. This event has put together the visual arts and the scenic arts, thus transforming themselves and their environment totally. The increasing importance of the communication could significantly transform our culture.

During the 20th century, the artists have proposed new points of view and explanations about the occurring changes. Sometime, it turned to be destructive because of their opposition to the ancient values, but some time it turned to be constructive when they revealed new possibilities and definitions. The role assigned to the artists by Marshall McLuhan is to be educator of our perceptions. This idea is borrowed from Ezra Pound, who called the artists "the antenna of the race".

This is precisely this faculty of perception that makes art a counterweight and a complement of science. The combination of them tend to complete the process of human discoveries. The artist work from the concept to the analysis, and the scientist works from the analysis to the concept. Science renews our knowledge of the world, and art renews the experience of the world. Implicitly, the artist communicates the scientific discoveries.

In the present time, art has changed its specific function. The functions traditionally attributed to art - intuition, revelation, catharsis, expression, representation, etc, - are currently far better exercised by the cinema, the television, the psychotherapy, etc. Art does not only interrogate our conceptions of Universe, but it interrogates the Universe itself. The social significance and the role of art should be seek in the process of communication (consisting in storing, treating and transmitting information). Art is no more concerning the production of pieces, either traditional mediums or electronics one- but well rather the discovery of the internal logic of a world view. Therefore, the contemporary artist works with traditional tools, ready-mades, media. He plays with his body, with the space, with the spectator and, finally, with the modern technologies.

Antiquity was a theatre era. In the Middle-Ages, the majority of people "read" the bible in the churches' stain glasses, and the feudal lords kept the memory of theirs ancestors through tapestries. The Renaissance had developed architecture and had directed the plastic art. The writing revived in the 19th century. Today, information and emotions are received through the screens of television or computer. The means of expression and the formats, imposed by these tools of communication, create specific languages. More and more, people get used to them. The artist using these new technologies would favor his message.

When the artist use techniques that create current consumption products, he benefits from their unspoken message. The artist does not need any more to express them. In the case when he uses old techniques, he would need much more energy to reach an equal complexity and profoundness. Therefore, using current techniques is much easier. Indeed, they already hold a large part of the present. "This is the reason why I am speaking of unconscious, If you use techniques that produce our current environment, They bring with them already so much reality, that it is far more efficient using them instead of the past one... Edmont Couchot [Musso 1993] establishes that there are two dimensions : the reality and the imaginary, and henceforth, a third one is appearing : the virtuality".

On the contrary, the new technologies need the artist to be deciphered and explored. Some new languages of communication, usable in the new space of communication - the virtual space - must be drawn up. This field of investigation, attractive for the artists, belong to their skills area. They already work on it.

Science and modern industry create more and more products. They don't know all the possibilities and destinations that this product occurs. As a consequence, a strong request of creativity exists in many fields of the new communication space exploration, as the new tools limits, or as the innovation in usual processes. The artists "guests in the house of the technology" use machines and " living systems as biological organisms, considering them as a material and searching how it could form". The experimental-artist Woody Vasulka [Housz 1995] gives a key allowing to explain the fascinating relation between the artist and the new technologies : "the poetic source of the 19th century was the nature - for romanticism or even impressionism. In the era of the urban culture, the technology has gradually created a environment sufficiently rich to provide us modalities replacing the nature's ones. The technology is the continuation of the nature, at least as the inspiriting source of my artistic work".

In the communication society, the artistic activity, as every others, must suit to the information traffic. The digital or digitalized images can be simultaneously everywhere and anywhere. The structure of the WWW expands on all the continents. Mail-art and Copy-art point out the importance of the transmission and of the information exchange. the screen-sharing work is a staging of information exchange in real time. The current exhibition is close to a "screening of virtuality, where emerges the diffuse and reticular organism of the creation" [Boissier 1991].

Between the different trends in art, that have led to Cyberart, two types can be distinguished : the artists who have preceded the contemporary Cyberart by exploring the relations between art and technology; and those who have developed the idea of an interactive art, of the spectator participation and of the collective creation.

The Industrial Art, the Maschinenstil and the Art Nouveau, at the end of the 19th century, can be considered as the ancestors of Cyberart. the Art Nouveau and its German variant, the Jugendstil led to the most deepened exploration of the problematic art-technology, and to its most resulted realizations at the end of last century. Thanks to its universal and eclectic stylistic language, the Art Nouveau has influenced all the forms of art - plastic or ornamental. This mixture of fine arts and decorative arts was based on an aestethic, that could be applied to non artistic objects and phenomenon. This mixture led to a radical simplification of fine elements, that gave to the Art Nouveau its high degree of abstraction.

At the beginning of the century, some other artistic movements, were the precursors of the Technological Art: the Futurism, the Dadaïsm and the Constructivism

the Futurisme didn't only exalt dynamism and speed, it endeavored to find abstract equivalent for all the forms and elements of Universe. The Futurism recommended to merge art and science. The futurists were the first to point out, in their ideology, the necessity to be associated to the modernity. So, theirs themes were associated with all the evocations of the new technologies at this time, as urbanism, transportation and movement speed. To break down the different phases of new technologies, the futurists followed the procedure of dynamic composition that conveys the ideas of speed and rhythm.

The Dadaists have questioned the idea of Masterpiece. At the end of the first world war, they wanted to point out the absurdity of the machine, to criticize the industrial civilization. However, Francis Picabia and many other artists, who participated to the dada movement, looked for new means to draw the modern life reality. Francis Picabia confided that his first arrival in the US was a revelation. Indeed, he discovered that the world genius was a machine, and that the art could find a very lively expression in the machine. Marcel Duchamp took the initiative to apply the machine aesthetic to a human being. He wrote that his pieces were not paintings but an organization of kinetic elements, an expression of time and space, obtained by the abstract presentation of movement. He reminded us that, when we consider a movement of a shape in the space at a precise time, we enter in the kingdom of geometry and mathematical, exactly as we do when we build a machine.

The dadaist have opened the access to new fine art procedures that questioned traditional codes. The ready-mades were the first, in Contemporary Art, to put off the object from the context. It was the sign of a different use of pieces in networks. The global work of dadaist had previewed many current components of Technological Art. For example, the automatic writing allows to generate unexpected relations between images.

Finally, the Russian constructivistes defended the idea that art was an autonomous and spiritual activity. Tatline launched the idea that an artist should have an engineer or technician formation to take place beside workers in the modern industrial society. His " Productivists " heirs took back and developed this idea.

At the same era, the purge of suprematism added to this process a decomposition of shapes and colors, that was explored later by artists working with the digital images. These different movements, characterized by the use of light and movement, led to "kinetics" and "luminous-kinetics" arts. They were among the main origins of Technological Art.

The start of this kind of art can be dated back to the years 1910th-1920th.. When Duchamp, Tatline, Gabo and Malevitch realised theirs first works utilising mechanical movements. At the same time, they were elaborating theories on this subject. This era was also marked by the appearance of the first mobiles created in the studios of Rodtchenko, Tatline and Man Ray. While the Bauhaus artists of the Weimar republic developed an art mixing light and movement.

The using of media did probably start with Moholy Nagy. In 1922, he ordered by telephone some enamelled paintings, to a metallic panel's factory. Using a drawing sketch on a squared paper, he dictated by telephone the positions of the drawings and the colors, to the factory product manager, who was noting the information on a similar page to execute these works. For Moholy Nagy, as for the dadaists, the technology was a symbol : the painting on easel was dead and the avant-garde art had to co-operate with the new technologies. However the real purpose of Moholy Nagy was to make a painting, the style was fixed in advance : a geometrical abstraction. Many critics have been expressed about this operation. They asserted that the telephone had a limited rule in this action and that it was only the mean to place an order, it didn't add much to the piece, except the slogan : " the artist designs, the worker caries out ". However, It was a kind of real time and remote interaction, because the communication tools interfered on the order, as well as on the worker's result. The painting would had been probably different, if it has been created in direct.

In the fifties, a cathode-ray tube was connected to a computer output. This system had been used the first time for a military purpose to supervise the American territory : the SAGE system. The digital image resulting was very simple and schematic. But it was a sign of a genuine dialogue between the user and the calculator, or in some way, a kind of interactivity. At the beginning of the sixties, Ivan Sutherland already showed the possibilities of the so called " interactive computer graphics ". But the images quality did not really improved before the seventies, when a new type of image appeared and was improved day by day. These images were in 3 dimension and moved. They were immediately used to military purpose. The screen didn't only display diagrams and graphic symbols, it opened to a new world. This new artificial and complex world was simultaneously virtual, realistic and abstract. From this time, the user was able to have a real dialogue with the computer, it was an unprecedented situation in the history of image.

At the beginning of the year seventy-five, the micro-computing development allowed non specialists to experiment the interactive image, using different ways. For example, the opened blank sheet of a word processor is a small but genuine virtual Universe. As Edmond Couchot [Couchot 1992] establishes it "we shall never forget, that there is no principle difference between a word processor, and the most sophisticated flight simulator. Both of them open the access to virtual realities who the user interacts and communicates with. Only the level of complexity is different. Actually, if computing is considered as setting to work simulating models, aiming to rebuild fragments of the reality and to act on them, then, computing is giving birth to a huge virtual universe. This universe integrates progressively the preservation, the duplication, the transmission and the production of knowledge and various information (images, texts, sounds, etc.) as well in scientific, technical or industrial researches, than in the communication and network area".

Beside the development of digital art technology, other basis of Cyberart can be distinguished. Franck Popper [Popper 1993] insists particularly on photography. According to him, the role of photography in the development of modernism and post-modernism, provided criteria, applicable to the role of digital media in the art area. Frank Popper develops an idea of Walter Benjamin, for whom photography played a key role in the renouncement of the art object notion as unique and original. Frank Popper showed how the neglected critical potential of photography, as a reproduction mean, was strongly used, in the sixties, by Andy Warhol and Robert Rauschenberg. The digital technologies of image - firstly the video and then the digital image - have accelerated the process, and aroused new changes. The evolution of these technology is similar to the photography one's - however faster and more direct. In fact, the photographic model, added to the disappearance of boundaries between art and non artistic communication system, "put down computer on the artist doorstep".

Copy Art was undoubtedly one of the origin of the Communication Art, at least in a technical point of view. Copy Art can either transmit existing information, as for instance a symbolic story or visual data. Or it can create new information, for instance by manipulating daily or poetic object in front of the machine. The artistic practice of Copy Art are extremely varied thanks to the large scale of photocopiers technical abilities, thanks to the various aesthetic objectives of the artists and thanks to the various ways for artist to act on the technical process. Two major aesthetic trends can be identified in Copy Art. The first one is, technically, very close to the photography process. It puts real objects and the machine together, without any further artist intervention. The second trend is closer to the printing techniques. It consists in transforming and combining images and shapes. The interaction process, between human being, creates the sense and, therefore, the culture.

Artist would prioritise system and process facilitating and amplifying interaction. So they can touch a world audience, using, for instance, a cable and satellite digital network. In 1977, Kit Galloway and Sherrie Rabinowitz attempted to connect together, by way of satellite, US east coast and west coast participants. This experience prefigured the opening in 1984 of the first Electronic Cafe. In 1980, Robert Adrian, associated to J.P. Sharp, proposed an event called Artbox. It was the first artist mail network. This system allowed artist to use their own telecommunication system. Later Artbox became Artex, the pioneer of artistic electronic network, internationally accessible. Artex was at the root of several telecommunication projects. The J.P. Sharp system is essentially a electronic mail service or a system of mail electronic, from where users can access to different localization.

Dario del Bufalo, in the catalogue of the 42nd Biennial of Venice (1986), gave his version of the technological art development, in the last half-century. According to him, the forties were the Electronic Art years, the fifties were the Data Art years, the sixties were the Uncertain Art years, the seventies were the Computer Graphic Art years, and the eighties were the 3D images years. To continue, we could call the nineties the Communication Art years.

In the seventies, the artists Op and kinetics, particularly Carlos Cruz Diez, the GRAV (group of visual art research) in Paris and the Gruppo T in Milan, tried to implicate further the spectator. They wanted him to realize the need, to use his liberty in a more constructive way. For instance, Carlos Cruz Diez work with the color's perception abilities. Some other French and Italian members of the group- particularly Morellet, Le Parc, and Colomboet Boriani - put the spectator creative abilities to the test. They put the spectator in unusual environments such as labyrinths or courses, punctuated with optical, stroboscopic, magnetic, kinetic and luminous effects.

The will to involve the spectator was obvious, in the happenings, actions, events and performances of the sixties and seventies. These realizations were usually critical of the society. In a same way, the graffiti and tag phenomenon is a violent challenge to the social established order. It was rediscovered by American artists in the sixties and largely spread in the seventies. Apart artistic circles, it allowed people from many levels of society to exert an aesthetic and creative activity. Even if usual social criteria consider it as destructive. The mural painting started in 1967 in the south districts of Chicago, with William Walker and an assistants team. The American "mural" movement had a double social impact. It allowed people of the district to become aware of their daily problems. And it led to the organization of groups that organized some collectives activities. The participants of those activities took part, in concrete terms, to artistic and socio-political expressions.

Some video installations implicate the spectator in environments made up with many images. the spectator is encouraged to produce his own editing, following his own pace and starting the way he wants. Or he is encouraged to associate images and elements, to travel in a space or to compose it mentally. Not only the eyes, but the whole body is solicited by the piece. The electronic on live, allows an other level of implication. The spectator body is captured, and is integrated, in real time, to the piece. Some time he can be moved in other places and other times. But more generally, the image is trapped, disappointing, escaping. It takes part to a critical inquiry on the representation and on its possibilities, on the relation between the bodies and the space, on self-consciousness, far beyond the "Participative Art" ("Interface" by P. Camus).

The majority of the Kinetic Art pieces, at these years, need the spectator intervention to be activated and to reveal their structure. They took a important part in the development of the concept of polysensorial spaces or environments. These spaces aim to produce psychological and perceptive effects, by using color, light and movement. The piece reacts to the visitor presence, to a gesture, to a movement speed, or to the body electric charge as for instance the "Interaction fields" of Piotr Kowalski or the "Cybernetic sculptures" of Tsai. The piece is activated by a switch. The connection between the cause and the effect is direct.

The temporal parameter was largely introduced in the pieces of the sixties. It opens the pieces to variation and hazard. It introduces the relativity of perception. And, it obliges to explore the pieces differently. The pieces realized by such artists as Robert Morris, Donald Judd, Richard Serra do not belong either to sculpture or to architecture, they would be later called installations or environments. This pieces, to be understood, need to be studied with many different points of view, in the time of an experience including both physical moving and mental memorization activity. They can be defined as a group among which the spectator is the main focus and where the spectator perception is appealed, is tested, and sometime, is the first purpose of the experience.

In the second half of the 20th century, and more particularly in the 60/70s, the artistic movement developed the idea of the proximity between the artist and the spectator in a common space and the idea of a shared responsibility between each other. They associated the spectator to rituals, and included him in the display. They used mirror interactions and electronic feedback, or they turned the spectator to the driving force of the action. And so, they contributed to the construction of Cyberart.

Contemporary Art appears as an art, seeking for a dialogue frame and dynamic exchanges with the spectator. For Edouardo Kac [Kac 1992] "the dynamic exchanges and the images, sounds and texts transformations, reflect more faithfully the computing era data flow, than the fixed forms". The art, that is built on a dialogue mode with computer, or rather, through computer, and that is utilising the cyberspace or existing in the cyberspace, is the Cyberart.

Cyberart has carried out the change from an ordinary participation of spectator to an interaction and to a collective creation. It has modified the relations between the process maker and the resulting piece, between the human intelligence and the artificial intelligence. It has modified the connections between simulation, virtuality and reality. The traditional artistic activity consists in representing the reality. But the cyber piece can turn to be the reality, as far as the cyberspace can be considered as a real space. In this case, the spectator is not simply a consumer in a mausoleum of objects. He travels and makes discovery. He takes part to the restructuring and the re-creation of the piece.

The Cyberart, as an art of dialogue and communication, is a part the aesthetic of communication research:

• Unlike traditional works, the creation is closely linked to the tool, because the protocol determines the visual qualities of the telematic piece. To create a virtual reality piece, the tool itself has to be created.

• Unlike traditional works, the creation is not a manipulation of aesthetic functions, but a manipulation of logical functions.

• Unlike traditional works, new working ways have appeared, particularly in the networks, as for example the analysis of data flow.

• Unlike traditional works, the development of interactivity leads to a collective work. This work often displays action on a world of shared meanings. It tends to elaborate a cultural identification and (or) a world consciousness.

• Unlike traditional works, the multiplication of remote-presence is not in contradiction with the cyberspace anonymity. A participation does not necessary involve a signature.

The Cyberart has permitted to re-introduce and to synthesize all the anterior problematic, to generate new ones and to display them as a communication action. The piece gets its reality from the observant / user active presence. Either the time is real or deferred, either the space is genuine or simulated, the image does not belong anymore to the reproduction area. But, the image thinks up its own world instead of reproducing our.

An artistic change, to lead to really new works, needs to touch the spectator in his privacy and in his essence. His imaginary, his language, his relations with others need to be touched. Techniques cannot only be used to ease the utilitarian tasks. It has to destroy some fiction, some dreamily world representation. It has to open new ways for human desires, to focus new kind of aesthetic creations. Techniques, even socially appropriate, needs this creative destruction power to be significant.
The Cyberart is itself a tool for the creation of the new language, coming from the global communication space and transforming it.

In Cyberart, the content or the interpretation of a piece is not modified. But the definition of the piece is. The pieces exercise an influence on the "immaterial", on the author himself, on his thought system. The pieces allow interaction with others' intelligence. But the author is among a network of interaction with systems and other authors, and so, he is no more really an author. The realized piece is not only immaterial, it also cannot be got, it is collective, indefinite ephemeral or temporary. The project ideas themselves are often dissolved in a computer program. The program operate alone at the realization step, but also sometime, at the conception step. It moves away the author and the piece. And, at the same time, it reveals all the steps from the initial invention to its realization. A new language, a new expression mean is necessary. It has to be more fluent, more symbolic. It has to allow several intellectual and temporal comprehension levels. It is impossible to create a musical piece, in a classical way, when it can be interrupt at any time. It is difficult to create an image, in a classical way, when its dimensions, its colors, its resolution depends from the computers protocols and capacities.

At the same time, the technology has made disappeared the references and the time and space constraints. It has allowed collective work in real time. This collective work implies that people with different cultures and languages collaborate. So, the elaboration of a common language is necessary. Some visual and symbolic references have to be unified.

The piece can be created automatically, it can be created in interaction with computer. This possibility is in keeping with a scientist and technical tradition of automaton and automatism. But, the new software of writing, reading, translation, switching and thinking, open further possibilities. Especially, when, in connection with artificial neuronal network, they simulate the human intelligence, and doing so, give access to the human intelligence basis and limits.

Georges Legrady established that, in the digital information process, as in other communication ways, the technologic elements - hardware and software - impose a format to the processed information. Although, they are usually considered as neutral and "devoid of proper values" [Legrady 1990].

The global communication space is universal and democratic. So an understandable language is requested for people non initiated to Contemporary Art. Fred Forest recalls " We shall not forget the power of such a number of creators, connected on the same network that benefits from the power of switching telematic. A group of 1000 people can create 1000 000 different interactions..." [Forest 1993]. This participants become co-author. The communication artists open for us new perception and action ways. On a poetic way, they make us become aware of a new space, abstract, enlarged by its nearly infinite switching possibilities, and becoming, day after day, our daily environment.

An illustration of the Cyberart nature is the artist attempt to involve the spectator in the creative process. Particularly, the way, that the artists used, to lead the spectator, from a simple invitation to intervene, to a more interactive and elaborated participation.

Franck Popper distinguishes "participation" and "interaction": "In the artistic context, "participation" has meant, since the sixties, an active spectator intervention, both intellectual and behavioral. This double invitation constitutes a breaking off with the traditional behavior toward the spectator. It has important socio-political implications. The spectator participation is not limited to a specifically artistic context. The events, where the spectators is invited, can often be compared to tribal or ritual feasts" [Popper 1993].

The "interaction" notion, has been more recently used in the artistic area. It gives to the spectator an even more important role. In this case, the artist tries to generate mutual exchanges between his work and the spectators. This process has became possible thanks to the recent technological systems. A situation has been created, where the art work reacts (or answer) to the actions (or questions) of the spectator/user. These realizations are usually global networks and they request a total implication from the spectator. Their meaning is undoubtedly sociological. But they focus on environmental and daily problems, rather than on political problems, as the 60s realizations did. Sometime, their way is clearly scientific. Therefore, in the Contemporary Art context, "participation" indicates the relation between a spectator and an accomplished art piece, while "interaction" indicates a mutual exchange between an user and an "intelligent" system.

Up to recently, particularly in the US, the interaction was only used to indicate the exchange between the artist and the user. Today, it is also used to indicate the relation between the artist and the spectator, connected through networks. This networks can be a very simple electric or electronic system, as well as a world-wide or local group of terminals. In this enlarged context, the creative activity is no more limited to the professionals : artist, architects, or composer. A wider audience is concerned.

The Cyberart purpose is not necessarily to create a piece. It can be also to experiment a piece. The experimentation is anyway the favorite field of Contemporary Art in the 20th century.

The development of Cyberart can be compared with the development of video, an other experimental art. At the beginning of the seventies, video was only a potential way of art. A few pieces had been created. A small number of visionary artists, as Nam June Paik, Peter Campus and Ed. Emshwiller, understood the video's possibilities and showed the way. Then, an increasing number of artist, exacting on their world vision, worked on video. And progressively video defined its distance with cinema and television., and became an artistic expression mean, in th same way as another. The Cyberart is reaching the same evolution stage. The 3D image specificity is not to reproduce the subtlety and the spectrum of painting colors. Its own structure and its unlimited transformation potential, are its best qualities.

Cyberartists can be divided in two groups. In the first one, the artists utilize cyberspace to obtain new fine art elements. In the second one, the artists turn the communicational experience (process and product) to a proper work. But poetry is still present. And the machine intervention is more or less important. In that case, the machine is not rejected, but its functions are used differently. The purpose can be either critical and accusing, or friendly and accomplice, toward the Machine.,Art is then appearing, rather, as a research and experimentation field.

The artists, working with telecommunication, move in a different context. Their works are more closer to conventional work. They are constituted with two-way explorations, and with dynamic relation between the process and the new technology.

This research has an experimental nature. It combines event and experience. And a piece can be resulted or not.. No hierarchy exists in the process and the artistic product. The process is really important. For Fred Forest, Communication Art is not only something to see. It rather allows us to take part to the production process. This can be apply to Cyberart when artists intervene directly and in real time in the field of communication reality, and when they propose their own standards as an alternative or their own value as an ethic. Artists are still producing their symbolic works. They contravene sometime to the usual network utilization. Doing so, they reveal the emotional contents that can be released by the experience of the presence and by the distant action. They reveal their implication on the ubiquity and simultaneity.

Two types of Cyberart can be distinguished. The first one consists of a dialogue between the artist and spectators or other artists, through the computer. Generally, it utilizes communication networks. And it explores investigation fields such as Sociological Art, collective creation, space transformation, real and virtual merging and combinations. The second type concerns the production of pieces in a dialogue mode with the machine, the exploration of the creative machine potential, the simulation and modeling of artificial life.

The cyberartists try to intervene aesthetically in the human spirit functioning. They try to stimulate the human spirit, challenging it. They try to increase the consciousness level. They try to encourage the intuitive knowledge, using conceptual process. They try to develop the mental imagery, and to stimulate the memory the reasoning and other mental process. The art of the thought is given a multi-sensorial creative dimension. Doing so, the artists endeavor to develop the imaginary aspect of human intelligence and, to transcend the opposition between the spirit and the machine in a artistic action. At first, these artists work on abstract information instead of concrete materials. They hope to allow the audience to access directly, with their senses or their spirits, to the origins and to the process of the creative activity. Meanwhile, the dominant aspects of Cyberart are those implicating, very specifically, the spectator. Either, the artists use the piece visual quality, and appeal to all the senses. This is particularly the case for pieces using virtual reality. Or, They use different ways, to imply the spectator at a physical and interactive level.

The temporal and spatial dimensions have a key role in most of Cyberart productions. They are often link to scientific discoveries that. the artists, deliberately or sometime subconsciously, integrate into their pieces. The artists try to generate a new conscious of time, both instant and duration. They use the potential of the feed back and of the temporal difference. And they try to generate a new space perception, by simulating the spatial imagination. Some works evoke an unlimited space, constantly in movement or give the impression to float in this space. Others connect distant places. The purpose of the artistic space utilization, or of the space without wall, is to bring out a new spatial sensitivity. Many works use the interface science-technology-nature, to turn the technology into a more ecological way, and to unify the natural and the artificial environment. The Cyberartists endeavor to develop the spectator emotional stage and mental faculties. Doing so, they use performances that are experimental psychological matters. For Cyberartists, technology is a tool, not only for the hands, but also for the mind. This tool is able to modify our mental process and to sharpen our sensitivity.

The Sociological Art had existed before the introduction of Cyberart in Contemporary Art. Since the sixties, it was present in the "mural " movement and in the "happening action" of the situationists. The use of multimedia communication network has broadened the Sociological Art development. The artists intervene in the existing network as terrestrially broadcast TV, satellites, radio-broadcasting, telephone transmission, The transmission become the origin of piece production, a kind of invisible ready made. The artist intervention summon the information transparency. The unquestionable universe of communication appears suddenly more unstable and critical. The purpose is to turn into visible, the invisibility of the network regime.

Aesthetic of communication is one of the trend of Sociological Art.
Derrick de Kerckhove and Mario Costa gave a boost to many artists, particularly to Fred Forest, Christian Sevette, Stephan Baron, Natan Karczmar, Robert Adrian. These artists have realized projects as part of the group of aesthetic of communication.

Costa wrote in 1983 the first document introducing the group's double goal. The group wants to elaborate an aesthetic and psycho-sociologic theory, connected to the new communication technologies. And it wants to connect artists and researchers, interested by the project, from every where [Costa 1991]. The group process is based on the capacity of these technologies, to modify our real continuum's experience, and to create events independent of the space limitations. According to Costa, aesthetic of communication heralds a new spiritual era, based on an unprecedented merging between arts, technology and science. The aesthetic of communication is an aestethic of real time events, able to visually bring closer together places, separated in the space.

In these kind of manifestation, the exchange content is less important than the used network and the exchange functional conditions. The aesthetic object is replaced by immaterial fields voltages and by energies, either vital and biological (mental muscular, affective), or artificial and mechanical (electrical, electronic). Our space time perception, based on the object is modified. The subject itself is getting transformed. It is no more defined by the rigid opposition between the self and the non-self. But, it takes part to a common energetic flow. Finally, the event generate a new phenomenology, based on a, distant or deferred, virtual presence, that evokes an inexpressible "sublime" of Kant.

In an article, published in the review Leonardo, Kerckhove examines the new spatial sensitivity coming from the communication arts. According to him, radio, telephone, and computers are "psycho-technologies". They are technological devices of our body, that have acquired a global significance. However, psychologically the western society is not enough advanced, to integrate, in its daily environment, this new feature. At a social or political level, our reactions, toward the planetary environment, is still dependent on obsolete, inadequate and inherited from the Renaissance, concepts. A new environment, has come from the computing network and the satellite connection. To understand and to integrate its complexity, we need a new spatial sensitivity. From this angle, the aesthetic of communication would organize the manner in which the communication technologies affect our spatial concepts and our psychological relations toward the space [DeKerckhove 1991].

According to Forest, the group of the aesthetic of communication, interrogates the technological complex environment of the industrial societies. And, it proposes critical and creative reforms, in a break with traditional solutions. Its works are at the heart of the changes, affecting the industrial technologies and the communication. Its research is focused on the new sensitivity, coming from the information fast exchange on long distance. Its artistic activity is characterized by a distant physical presence, by the telescoping of the immediate with the deferred, by the interactivity free play, by the combination between the real time and the memory, by the planetary communication, and by the social group study. The communication artists want to emphasize the wonderful coming prospects, that are allowed by an inventive use of these new tools.

The artist thought is rather stimulated by the underlying networks symbolic order, than by their technical skills. Indeed, they allow the participant to access to new kind of space, to a new perception of time, and to a new area for imaginary.

For Katsuhiro Yamagushi, as for Nam June Paik, the electronic means has the potential to increase the user creative and intellectual abilities. He called his main concept "imaginarium", It was a users network. Using a video camera and a software, every body was able to film and transfer his film on the network. The images can be transferred on real time to every connected screen.

This concept is close to concept of the electronic cafe, or of the Renga. But Yamagushi prioritized the exchanges with the spectators rather than with the artists. This idea appears to be a good prospect.

More and more, the spectator is put on stage.

The important pieces, of the pioneer Dan Graham, illustrate the dilemma of many video artists. For Benjamin Buchloh, the piece of Dan Graham show that, from the beginning, on a technical point of view, the mass TV main discourse is at the origin of every video works, and is finally integrated to it. A Graham piece, realized in1971, "Project for local cables TV " illustrates perfectly this point of view. In that project, Graham's usual way to catch the dynamic exchange between two people, is linked to collective audience, through cable television. In the same idea, Graham's most perfect piece, realised with Dara Birnbaum in 1978, was "local Television News Program Analysis for public Access Television". This project is based on the idea that a direct public access to television broadcast is a matter of time and organization. And, so, TV will stop to be a social manipulation tool. It will become a tool of self determination, mutual communication, exchange and practice. Video Art takes advantage of the monitors network possibilities. It plays on the observer/observant relationship, that is to say the relationship between the spectator and the piece. It plays on the reflection of images in mirrors. It plays on the relationship between space and time. The installation of video screens and sculptures in a common space, creates a place where reality and fiction merge. In Dan Graham installations, the spectator is trapped by his own reflect. In the project "present continuous Past", a camera films a facing wall, covered with a big mirror. A monitor is put under the camera and broadcasts the film. The system of mirror and camera issue an innumerable number of spectator images, only limited by the screen definition. But, the video image is postponed for a few seconds at the beginning, and then the duration increases, so virtually, the spectator image never leaves the installation.

Interactivity between several persons and collective creation, are becoming the main purpose of artistic project. The artists do not deal anymore with the problem of image aesthetic. They use telematic networks to elaborate new ways of transversal flow, rhizomatic, that would work at a different scale.

The communication process is put on stage.
The project of Steven Wilson, "Parade of Shame" (1985), is probably the first precursor of interactive television. It is a series of 8 animated films on computer. It deals with the biological and ecological evolution. The spectators of San Francisco, equipped with TV cable, were able to intervene in the event. The main theme was the extinction of some species. The spectators were connected by telephone to the broadcaster, and they indicated their choice, using phone keys combinations. "The Venus of Willendorf in Synthetic Speech Theatre" (1985) was presented, the same year, in the exhibition CADRE, in San Jose, California. The public was able to intervene on a debate between four virtual people. The virtual people had been created by computer, and were shown on a digital screen. They used synthetic voice system to speak, and understood the spectators questions thanks to a speech recognition system. These systems, created by artists, were adapted to commercial purposes. They now appear on most of the softwares.

Loren Carpenter's project "Kinetic Evolution" was a performance. It involved the participation of several hundreds people. This piece aimed to activate, to get, and to explore the creative collective potential "The group intelligence". It took place in Siggraphs, Imagina, and Arts Electronica 94, on a specific entertaining way. Bicoloured reflector panels, the " Wands ", were given to the public. The simple signals, broadcast by the Wangs, were picked up by a computer, and generated the animation of some simple geometric shapes, on a giant screen.

Brian Reffin Smith realized in 1988 the piece "Artist/critic". It was composed with two computers, connected by a "clever interface", the public. The first computer allowed to draw vertical lines. The second computer allowed to draw horizontals lines. Their combination allowed to draw diagonals and curves. To realize complex drawings, the two participants had to find agreements. Like that, their communication process was exhibited.

People have to learn to communicate. Art in shared space can be used to.
Artists protest again the passivity of the spectator, against the solitude and selfishness of the contemporary society members. They try to teach people to communicate again. The electronic systems allow many artistic experiences in the communication fields.

Already in 1977, two communication artists, Kit Galloway and Sherrie Rabinowitz, realized the first interactive dance performance. They had gathered, participants from US Atlantic coasts and Pacific coasts, in the Goddard NASA Space Flight Center, Maryland, and in the Educational Television Center, Menlo Park, California.. And they transmitted by satellites digitalized images, to the both places.

In 1980, Galloway and Rabinowitz, associated with the mobile image group, realized the project entitled "Hole in Space". This project included a satellite connection between New York and Los Angeles. Cameras and video monitors had been put in the shop showcases of these two towns. So, the publics of the two towns, were able to communicate visually and orally. No advertising had been done, and no indication or instructions had been given on place. The people discovered " Hole in space " by chance, attracted by the screen images. A space-time hole was created, and the crowd of the two towns communicated within it.

In 1984, Galloway and Rabinowitz created their Electronic cafe (CAFE = Communication Access For Everybody) at the Contemporary Art Museum of Los Angeles. Six distinct cultural communities of Los Angeles were connected to an image bank, a telecommunication network, and a data base. Television equipment, graphic tables, computer terminals, printers, cameras, video screens and a bulletin board had been installed in the museum and in the 4 cafe taking part to the project. The consumers exchanged theirs images, used the camera for artistic actions, or wrote poems. So that, the project created an open " wall-less " gallery, where the public took part to an artistic communication work.

Since the Electronic cafe, many art experiences have been realized in shared space. As for instance Documenta in Kassel (1992), the Biennial of Venice (1993) and in other festivals in Denmark, in France (Paris, Toulouse, Nice) and in Finland (Helsinki)

In January 1995, a jazz concert took place in the shared space. It gathered simultaneously the groups "Dictateur du vent" in the gallery Natkin-Berta in Paris, and "Pauline Oliveros and the deep Listening Band" in The Kitchen in New-York. They were connected by a Numeris connection. During this experience the participants really felt the fusion between the humid and windy Parisian night and the sunny New Yorker afternoon.

Beside the technical aspects, these experiences arise many questions about the new delimited virtual space. The network becomes an integral part of the shared space. The audio-visual time disrupts the live show time.

The cafe is based on the scheme of an international opened space, where the artists guide the public. It allows the participants to explore themselves the new evolving space of interactive communications. It also sponsors artistic events, in all disciplines : musical interactive performance, dance, concerts, conferences, on line chatting. Several web sites are usually involved [Galloway].

Another example of artistic activity, in the shared space, is the project "Artists network". It was created in 1992 by Don Foresta and Georges Albert Kisfaludi, with French and foreign art schools. Infographic tools and video equipment were given to the students. And they exchanged their works through Numeris network and works together on the same images. "Artists network" was present in 1994 in 14 towns, in 4 countries : France, Germany, US and Japan.

The international electronic cafe, and "Artists network have in common the artists. They are the intermediaries between the public and the technology. And they help the participants to interact within complex new technologies devices.

"Conversations with Angels" is a multimedia art work on Internet created by Andy Best and Merja Puustinen employing 3D computer graphics. It combines visual environments, animation, video, sound effects and texts. The multi-user capability accessible via the Internet creates a whole new status of existence - an art work as a social process, allowing the construction of an online community within the artwork. In the worlds of this web project the viewer can meet other visitors by using an "avatar", a 3D representation of himself in the virtual space. "Conversations with Angels" is a platform for communication. The multimedia art project is carried out as a social process and provocateur. Besides being entertaining, it also encourages the viewer to explore, investigate, test and question preconceptions about the nature of the Internet and virtual spaces. The visitor to the world is first taken on an animation which sets the tone for the narrative. The user navigates themselves into an aeroplane, from which they emerge into a psychedelic courtyard. The yard acts as a "hub" for a series of individual spaces inhabited by "bot" characters such as Carl the serial killer and his bored wife Marg, Anne a lesbian princess who grows plants, and Bob the red neck goldfish…

Cyberart shows currently two ways to transform the space. The first one, consist in giving the real space, a cyber aspect, to make it interact with the spectator. The second way, consists in making real the cyberspace, and in establishing relations between people or distant events.

Architecture is a communication system in Cyberart.

Virtual Reality as a new medium requires a genuine and specific formal language. "Staging strategies" by Lilian Juechtern and Nicole Martin is a project in virtual space. The research on basic parameters for the design of a Virtual Reality has thus been constituting for their work. The basic element of a virtual world is dynamic information. Events, data, everything is floating and moving. Dataworlds lack gravity: there is no Up and Down, no ground, no horizon. One move and everything is changing. How can we find our way in a Virtual Reality? To investigate this question, "staging strategies" focuses on the design of an abstract Virtual Reality written in VRML.

Christian Moller builds in his works, virtual or half-virtual, architectural environment. They are based on the interaction between different data, coming from the real world. He interpreters the building languages with a metaphorical point of view. His installations utilize electronic devices for an architectonic conception. They introduce us to a new architectural kind, based on the notion of dialogue. The virtual interactive architecture of Moller reacts and acts. It has literally a dialogue with spectators. In such pieces as "Electronic Mirror" (1993), "The Virtual cage" (1993) and "Kinetic Light sculpture" (1992), the spectator becomes the main actor. the installation centre point. Architecture is no more considered as an object of perception., but as a communication system.

"Road of light" (1994-1995) from Philippe Rouanet [CNAM], is built with 2 dimensions : the memory of the place and the urban frame of the place. "The flat country" (1994), from the same author, explores an other frame : the communication network, and explore an other place : the space between two worlds, the interface. In "The flat country" two identical installations are located in two different places, and interact in real time. The interaction displays the experience of our relation with the piece, placed between us and the other. When a visitor is walking in the first installation, he provokes light waves on the second installation grounds, and vice versa. These lights make each visitor feel the invisible presence of the other. Also, a dialogue takes place between them. The entire visitor body is involved in the piece. Every step is comparable with a fall ahead. It waits the light on the ground before moving. This space is ruled by the light that symbolized for Rouanet, the material of the visibility. And Therefore, the space is ruled by the view.

Artists work on the reality retransmission.
"Telematische Sculptur III" (1994) of Richard Kriesche is an interactive sculpture; It shows an original point of view on the relation between virtuality and reality. A rail, symbol of the physical strength, moves on rollers. A computer, whose screen is facing the rail, activate the rail. The rail moves forward as far as it crashed into the screen..

"What is the weather in Caplan" (1993) from Bernard Gortais, is a representation of meteorological data, recuperated from Caplan in Britain and transmitted to the exhibition place. The sent data modify the colors and the image composition, displayed on a computer screen.

Fascinated by the light and by the fluidity, Jake has a project "World utopian process". As every artistic creations, "World utopian process" starts with a frame. A specific technical intervention, composed with a combination of luminous phenomenon processing equipment, is going to transform it. This Frame is nothing else but the Universe. Indeed, the luminous phenomenon will be caught in the place of intensive life, in the main capital in the world, and it will create the piece. In a technical point of view, "World utopian process" is a system that catches, transforms and transmits the material reality, revealed by the light. The number of installation points can change. They are connected by satellite, Internet or any other communication means. At first, the light information, dynamic or static, with variable speed and intensity, are picked up. In real time. This expanding materials are extremely compressed, up to create an icon "voxel light". This icons are in a perpetual mutation. They recombines a reality, inaccessible to the perception otherwise. The created icon are transmitted to different points in the world. In the spectator point of view, "World utopian process" offer a kind of "Fountains of images", a global capture, totally utopian of the planet.

In September 1989, in a performance called "lines", Stephane Baron drove a car along the meridian of Greenwich. Sylvia Hansman accompanied him. They sent, regularly, images and texts in relation with their travel, to height correspondents, living in different European countries. This utilization of a instantaneous communication technology constituted a new representation of the line. The line is a fundamental symbolic sign. The faxes were the fractal projections of the meridian. Otherwise, every line fragment was a representation of the complete line. The perception of the meridian gave the impression to the participants, that they were measuring it.

The merging between the reality and the virtuality is an extremely important direction of the Cyberart development. When the spectator and the artist act on the real space, they can modify the virtual space data. The two heterogeneous spaces mix together, exchange elements, establish a dialogue, confront one to the other, mix contradictory perception. The reality hybridise itself with virtuality. The body logs on to the machine.

The interface is put on stage.
The interface role is essential in the works, exploring the merging between the reality and the virtuality. It is even the main concept of some works. The term interface has a wide sense. It can designate as well hypermedia browser, than the numerous sensors and prosthesis giving access to the virtual realities. The interface choice can have radical consequences. As for instance, the use of visualization glasses, hides the environment. And therefore, the image notion disappears for the stage notion. On another side, the trend is to hide the captors connecting reality and virtuality. The impression of a direct action of the body becomes the favorite interface. It is favored to the interpretation. The spectator feels less as an operator, and more as a magician. Also, his eyes can destroy what he is gazing at. And when he stroke an alive plant, he can engender, on a screen, the simulation of the plant growth, as in the piece of Christa Sommerer, Laurent Mignonneau and of Karl Slims.

"Galeria Virtual" of Roc Pares is a research and experimental production team, working on interdisciplinary projects which integrate Contemporary Art and digital audio-visual technologies, paying special attention to Virtual Reality. The main interests of "Galeria Virtual" may be found in the study of the specific properties of information technologies and their impact on contemporary culture. The methodology of the team is clearly interdisciplinary. In each of its projects, "Galeria Virtual" designs and develops new interactive systems and interfaces which redefine the concepts of space, time, light, body, subject, object, gender or limit. "Galeria Virtual" is both a conceptual and technological platform.

Agnes Hegedus has invented an original interface for "Handsight" (1992). A bowl, representing a big eyeball, has a sensor. It is hold in the hand. And it has to be immersed, in a transparent Plexiglas sphere, to explore the data base. The process is ingenious and aesthetic. But its metaphorical dimension (the eye and the hand), and its paradoxical dimension (the endoscopic investigation of the transparency) are even more interesting. It puts in resonance two memories, two spaces. The first space is the content of the bottle, a kind of Hungarian popular tradition of ex-voto. The second space faces the first one. It is a virtual image displayed on a circular screen. And it refers to the elements of the bottle. Between them, the interface, the junction between the two globes, doesn't give any visible information.

A new kind of performance, the art of the body, is launched again by the connections of the body with electronic and computing systems. Also, Stelarc [Stelarc] put on stage his body. He covers it with sensors and prosthesis, particularly, an artificial third arm. His breathing, his muscles voltage, and his heart palpitations produce a certain number of effects. And they define the angle changes of the cameras that film Sterlac in live.

"Liquid meditation" by Margaret is a narrative virtual environment based on nature, philosophy and architecture. Within the world, an interactant can explore meditative abstractions of natural water reflections. Exploration occurs within an upward winding architecture that structures the narrative philosophy, as well as the images of reflection. As a participant journeys through the structure, meditative experiences within the environment foretell the upcoming revelation. Capacity to achieve awareness of the philosophy is solely based on the skills and patience of the interactant. Within the environment, philosophy is structured as a visual experience. As in life, interactants within the virtual world explore and navigate of their own accord. The world exists in the form of both logical and chaotic experiences. These experiences exist within the world despite the path of the individual. Encounters with the unexpected may trigger decision-making responses. Persistence and dedication to exploration allow an interactant to reach higher levels of aspiration. In reality, virtual reality is a medium that allows visualisation of abstract concepts, as well as commentary on our existence. Within "Liquid Meditation", the focus tangibly is on nature and abstractly on philosophy. By exploring these elements of reality in an abstract manner, perhaps new awakenings can occur in regard to their presence in the world.

Some artists merge the reality and the virtuality by inter-displaying the spaces.
To prove the space fluidity, Jake, either places virtual objects in real space, or turns real object to virtual by displaying them. Her work "virtual fluid" (1994) is an installation. Some plane images are put in a three-dimensional space. They are produced on video screen, or are captured by the system itself from the reality around the piece.

The plane images, used, have different origins: cinema, photography, or other graphic elements. They are filmed on a video camera, recorded, and then broadcast on a television screens. They also can be replaced by visual events, choreographic actions or object animations.

In "virtual fluid" The observer is put in a fluid relation toward the time. The present, the past and the future appears simultaneously to him : The present of the images that he can look at, the past of the recorded images and the future resulting from his own behavior. A similar virtuality of the space is imposed to the observer. All Euclidean and classical perspective references are upset by the inversions, anamorphosis, scales changes, reflections, superposition and transparencies, allowed by the virtual screen.

One of the Cyberart aspects is the "Dataism"[1]. This art is opposed to the iconoclast features of Dadaïsm. By his formal practices, the Dataism reaffirms the traditional aesthetic. The "Dataist" works are not only unique art objects. They are algorithmic processes and data bases. The Dataist art works can appear as existing in a 3D space, and as moving in the time. But, they can also remain entirely synthetic and become only imaginary products.

The cyber artists put pieces connected to interfaces, at the spectator's disposal. The spectator builds his own piece, by exploring the data base. This is not a collective creation, because there are no real exchanges between the artist and the spectator. It is rather an evolving piece. It adjusts itself to the state of mind and to the personality of the spectator. Three means to practice dataism can be distinguished : the data bases, the interactive film and the universe's creation.

"Grammaton" by Mark Amerika is a "public domain narrative environment" created for the World Wide Web. It is presently made of over 1100 text spaces and 2000 links, a unique hyperlink structures by way of specially-coded Java scripts, a virtual gallery featuring scores of animated and still life images, and more story world development than any other narrative created exclusively for the Web. A story about cyberspace, Cabala mysticism, digicash paracurrencies and the evolution of virtual sex in a society afraid to go outside and get in touch with its own nature, "Grammaton" depicts a near-future world where stories are no longer conceived for book production but are instead created for a more immersive networked-narrative environment that, taking place on the Net, calls into question how a narrative is composed, published and distributed in the age of digital dissemination.

"dwtks" is a project of Knowbotic Research [Knowbotic], a group of researchers from the Kunstshochschule fur Medien art. "dwtks" is connected with the idea of associative memory. It refers to the portrays of "The four continents" (1764-1766) by the Antwerp painter Jan van Kessel. "dwtks" the contemporary outline of "Antarctica", one missing, undiscovered continent in the manneristic series of van Kessel.

The "dwtks" project is based on the information sets provided by current Antarctic research in data networks. This public appearance of scientifically collected and processed data series, mock ups and simulation reconstructs system effects of natural processes and thus produce a Computer Aided Nature. The data presence of an "artificial" reality. Knowbotic Research evokes in this a Computer Aided Nature interactivities between the navigation of a visitor within a self organizing agents-system. the emerging matrix of correlating energy transform the information territories into a data landscape.

The information territory, a self regulative data space, is outlined in dwtks by mobile unites. These multidimensional data bodies, labeled by Knowbotic Research as knowbots, are enabled to incorporate data energies of very different qualities and disciplines. The knowbots in "dwtks" embody the strategies which generate the dialogue on a hypothetical nature. These focused data stream are continuously

[1] determined by de Witt.

updated by its contact to automates and robotics measuring systems in the reference nature.

This knowbotic nature recombines natural system correlation with specific social, cultural and aesthetic components. this chimerical nature, unlike hermetically sealed virtual world, owes its existence to its linked effectiveness to the reality.

David has created "Waxweb" the network version of his first feature "Wax or the discovery of television among bees" [Waxweb]. Waxweb was perhaps the largest integrated narrative database on the world wide web. "Waxbeb" consists of 560 clips MPEG (an entire feature length film), 2240 audio clips, 1600 stills, and 900 pages of hypertexts with more than 5000 links. A custom interface allows the reader to add immediately visible hypermedia to the document..

Another example of interactive film is "Utopia" (1993) by Max Almy and Terry Yarbrow [Utopia]. Utopia is a multimedia installation which explores the state of the environment and specifically the crisis of the contemporary urban environment. The core of the piece is an unusual interactive video game hosted by performance artist, Rachel Rosenthal, in which the viewer is asked to make a series o choice heaven/hell, power/impotence, fiction/reality, utopia/dystopia...

The town planner and the contemporary artists often display their universe as towns.
"City Project" (1989) by Matt Mullican is a walk in a imaginary town. From an aerial view, this town looks like a base-ball ground, divided in clearly delimited zones, with different colors. The red zone is called "the Subjective", and symbolize the spirit. The white and black zones, are called " signs zones ". They symbolize a domain where the language does not exist under the shape of signs and symbols. The yellow zone corresponds to the "enclosed world", it is a microcosm symbolizing the entire world. The blue zone is, the "non-enclosed world ", and is closer to our living world. Finally, the green zone, "strengths elementary", symbolize the nature and the material. When a walker enters into a specific zone, the entire town take the color of this zone. It is to symbolize the abolition of geographical borders for a global social commitment of the town.

The bicycle can be another mean to journey in a virtual universe. "The readable town"(1989) by Jeffrey Schaw is an infographic interactive installation. The visitor can ride a bicycle in a simulated town representation. This virtual town is made with 3D letters, generated by a computer. Key words and sentences are built along the streets. The plan of the city is based on the map of Manhattan, Amsterdam and Karlsruhe. The architecture has been entirely replaced by a new architecture based on letters and texts. Therefore, a journey on bicycle in the town is a reading travel. Going on a way, is choosing texts and their spontaneous juxtapositions. The town identity becomes the conjunction of senses generated by the words, during a completely free journey, in the virtual urban town.

Another cyberart movement is "Biological work".
The Cyberartists explore the interactive possibilities of the human bodies, either to discover the invisible qualities of the participant subjects or objects, or to establish

symbolic relations between the people behavior and theirs visual or sonorous displays.

Also some artist work on the Movements capture.
"Doppler Dance, a sculpture of Waves", the performance of Steve Mann, is composed with six "rooms", each one including one or several radar placed around a stage. These radar transmit waves non discernible by the dancers. They interpret the dancers movements into different luminous or acoustic signals. Also, the dancers are in interaction with the rooms, simultaneously (the radar can see through the walls) or successively (by moving from a room to the other). When they move, the dancers explore the invisible sculpture constituted by the waves. The waves sculpture can only be seen when they are coupled with sonorous or luminous waves. The sounds are intimately linked to the human body. More or less dissonant, they reflect every movements captured by the radar. The radar does not control other music instruments. It is itself a instrument.

Michel Redolfi is a creator of sub-aquatic concerts. His work, "In Corpus", realized in Toulouse in 1994, during FAUST, was a cybernetic show. The immersed audience explored a kind of collective dream, composed with sounds and lights, influenced by his own movements. The participants carried a simple colored swimming cap, detected and followed by digital video cameras. The captured information introduced shade, live, into the audio-visual concert production. The result is different in every concert. It is the reflection of the particular character of the different auditor-performer groups. " In Corpus " promote an inventive sensoriality that express itself in a 3D, real and not simulated, cyberspace.

In "Travel to the garden of delights center "(1986) by Jeffrey Schaw, the images change as the spectator moves to the projection screen. The changes are activated by infrared captors placed along the way, followed by the visitor. In this installation, the dialectic between the figurative image and its digital structure is organized as a sequence of three images changes. The first one is a collage in relation with a work realized by Yves Klein. The second is an extract of an Oshima's film. The third one is a filmed extract of the Bosch "Garden of delights". The image changes from a high-colors-resolution to a low-colors-resolution. The Bosch image progressively emerges from an abstract raster, to become perfectly clear. Also, the travel in the smallest image detail allows to reveal a high level representation of another image.

The image-virtual space created by these works intended to coexist inside the physical-real space of the exhibition hall. This hall was composed with seven zones, delimited by the Gothic vaults of the museum architecture. Therefore, the move, from an image to another, was a kind of virtual journey, through the exhibition space. The images sequences were virtually placed along the way followed by the visitor

"The winds door" (1995) by Didier Lechenne and Daniel Pressnitzer [CNAM], is a visual and sonorous interactive installation. In where, every visitor movements release a real time interaction, with a displayed image and a propagated sound. "The winds door" put on stage, on the one hand, a reversible and non linear time, and on the other hand, a memory, considered as a stratum, a level of the past, to be rediscovered and updated.

The matter here is to used the spectator blow as an interaction. Unlike the previous example, the spectators act consciously. The movement, as well as the result, are natural, but they are reproduced by the computer. Pieces, by Edmont Couchot, as "The feather", "I sow to the four winds", or "The sail" illustrate it perfectly.

In "the feather", the spectators blow modify a three-dimensional image. "I sow to the four winds" is an impressive demonstration of interactive simulation. It opens, completely new ways to the computing artistic utilization. Edmond Couchot explains " the digital breaks up the image into its ultimate components: the pixels. But, while this breaking up makes the image, at least theoretically, unchanging, infinitely duplicating, transmittable without loss; and so stable, fixed, conform to the traditional image properties (painting, photography, cinema, television). At the same time, it gives to it the fluidity, of the numbers and languages, the capacity to answer to the slightest and most unexpected prompting of the observer. And so, the image becomes unstable, mobile, changeable, understandable. Henceforth, the image life can depend only on a blow. But, this blow sowed, to the four winds, broken fragments of the image surface. Also, the image takes, from the blow, the power to rebirth elsewhere, or, finally, to be more than a simple image".

Some artists attempt to establish a kind of dialogue, directly with the machine.
They explore the creative potential of the computer itself. They creates software that the computer has to respect, but, that does not prevent the final result to be relatively unforeseeable. In this point of view, the machine stops to be a simple tool, to become a creator, or at least, a stimulator of the artist's memory, reasoning, and brain.

Maurice Benayoun has invented his personages, by following the digital logic. His Quarx (1993) are not the copy of existing animals. They are based on algorithms, used to modelize 3D shapes. The metaphor is clear : if the world becomes virtual, we are Quarx.

With "Is God flat ?" (1993) and "Is Devil curved ?" (1994), M. Benayoun puts the spectator in the virtual reality. He abandons the principle according to which a virtual image is a set of geometrical models, in a set of physical models. He do not seek anymore to reproduce the world surface, but the laws ruling it. The realism turns to be an analogical modality of the world representation. The analogy is present in the process, in the generative components of the reality. " Before I believe the world, it would have to believe in me". In "Is God flat ?" and "Is Devil curved ?" the image literally feels the spectator presence.

The Mario Canali's "Virtual cover #3" illustrious his original idea of Cyberart. Following Mario Canali, the computer is not similar to the mind, but to a body. As a body, it is a part of reality. As a body, it produces reality. As a body, it talks to us with the language of perception, not with the language of mind. Studying the computer, we found that the matter is able to produce ideas by itself. It produces ideas, but these are not abstract ideas, but body-ideas. We can call these body-ideas archetypes. The virtual reality generated by this new body is the space of archetypes, a new space for art.

The modification and simulation of the perception is an important trend followed by some artists.

Within theirs interactive games, Nancy Burson, Ed Tannenbaum and Jane Veeder put the spectator in relation with the machine. They develop an endless loop feed-back between the spectator, the camera and the video monitor. In such work as "Composite machine" (1988-1989) by Burson, the artists added a image digital process, allowing the spectator to modify his own electronic reflection. "Simulation" (1986-1988) by Tannenbaum, allows to create new symmetrical faces, indeed, it doubles the right or left face of the spectator. "Warp it Out" (1982) by Veeder, allows the spectator to widen and to distort his face, and to add to his face some symbols and or pattern. Also, the spectator perception is modified, the biological criterions are replaced by electronic ones.

"Detour", the Rita Addison's virtual work, explores sciences, medicine and art through the autobiographical story of the artist. Rita Addison sustained a brain injury in a 1992 car accident that changed her life and her art. To create the virtual environment she uses her painting deformed by a special software who simulates the perception of mentally sick persons.

Rita Addison believes, the virtual reality can be a powerful way to evoke and stretch empathy capabilities; to share perceptual phenomenologies. She wants to create virtual environments which can empower us all, no matter what our abilities or disabilities. She wants to create accessible meaningful experiences of wonder.

Since 1991 the artistic researches of Isabelle Levenez are based on the studies of persons suffering from a physical or mental handicap, particularly on the studies of their hidden dimension, of their territories, and of the space necessary to their stability. Her installation, "Tetanization", is a sonorous environment. It explores the eyes conditioning when the lights are off. It trains the hearing and the touch in a narrow space. Isabelle Levenez puts the spectator in an empty space. In this place, with different mobility and recognition sensory, the spectator is confront to another reality, the blind's one. This sonorous space is a concrete architecture, composed with four cubes, spaced out with corridors. The spectator is invited in this nocturnal space, to circulate and to walk. A sonorous space is built, determined by the spectator touch along his route. The sounds guide the spectator. They give him a perception of the corridor dimensions. The spectator presence provokes a sonorous wave. The power of this wave decreases, as the spectator moves away the place where it was generated. The sonorous wave ends when it meets the wall, facing the exit.

Computer-aided "analysis" is a trend only usable in Cyberart.

The Richard Kriesche's installations are the "virtual data spaces" representing the modern information age as a universal waiting room. The sculpture defines itself purely on the basis of digital information traffic. It no longer makes any difference, whether this is artistic or everyday data traffic; Rather the symbolic signs of the computer in the virtual data room are once again symbolically loaded : the material rail didn't carry anything anymore, on the contrary - it's now the non-material symbols which carry the rail...

The movement of the "Telematische sculpture III" was controlled not only by the data relating to the art work, but also by the whole internal data traffic of the exhibition building itself such as telephone, fax, telex etc. The momentary condition of the sculpture was visible in the form of the status indicator on the screen. Every information input resulted in a visible change, the non-material symbols.

The simulation and the modeling of the artificial life has an important place in Cyberart. Currently, we assist to the elaboration of techniques, simulating the perceptions and the behavior of human people. Soon, our total knowledge could be apply, directly or not, to develop our aesthetic environment. New performances, exceeding the real environment limits, and using no reality element, can be imagined.

The theoreticians, for whom the simulation concept is the center point of their researches, tackle the problem with very different point of view. Jean Baudrillard assimilates the simulation to a "desertion of the real". For him, the abstraction associated to the cartography, to the double to the mirror, traditionally needs before, a real referent. Philippe Queau has a completely opposite position. He considers the simulation as a writing tool, opening a new territory to the creation and to the knowledge. For Florian Rotzer, the computer-aided simulation, results from perceptions, transferred from the human body to machines equipped with sensory captors. Also, the simulation could result from a total programming of the environment.

The natural selection can be controlled by the artists.

"Art Evolution program" by Wiliam Latham, was created in collaboration with the scientists Stephen Todd and Peter Quarendon. It allows to make a kind of natural selection, controlled by the artist. With this software, the artist decides, on a aesthetic point of view, which form of life will survive and which one have to die. The live forms, generated by the software, have an extremely organic appearance. Using this technique, the artist can explore a multidimensional space with millions sculptural forms. Then, the images are photographed and filmed. The resulting piece is composed by these pictures, of non-existing organisms, produced by a computing evolution process.

"Feeping creatures" of Australian artist Rodney Berry is an interactive virtual environment (cf. chapter 9). It consists of a population of artificial organisms called feeps. Music and video are continually produced through the feeps' evolution and interactions. It is driven by an aesthetic based on systems and processes rather than objects and images. Nature-inspired processes develop a diversity beyond the direct intentions of the artist. The artist is offered a chance to step out of the role of God/Creator to become more a farmer or hunter-gatherer of aesthetic experience. The feeps' world is a flat green grid across which the feeps and the camera move.

The Participant moves a mouse to steer a virtual camera and microphone across the grid, causing the sound and images to change. Each feep has a sequence of musical pitches forming its chromosome. Feeps choose their mates by comparing musical intervals, then portions of each parent's note list are passed on to the offspring to form a new chromosome or feepsong. Rhythm is passed through the food chain. The result is a continually changing web of inter-layered musical cells, an evolving sonic meta-nature.

The concepts visualization is very popular among cyberartists.
"Carla's Island" (1982-1983) by Nelson L. Max is a three-dimensional digital animated image, produced in real time by a computer. Its parameters can be modified by the public. Its purpose is not only to provide a visual pleasure, but above all, to obtain, by this computing simulations, a near photographic realism, allowing to visualize scientific and mathematical concepts. Thanks to an analogical complex system, the spectator can act on a three-dimensional moving image, representing island surrounded by water. He can also modify the colors of the sky and the sea, and the position of the sun and the moon from the horizon.

Some artist work on the modeling of the life.
The work of Crista Sommerer and Laurent Mignonneau explores interactivity, virtual reality, communication and artificial life. "Interactive Plant Growning" (1992) gathers a cluster of three-dimensional images generated in real time and projected on a screen. The roots of different plants species are fitted with electrodes which transmit electrical currents from the plants to the computer. Converted into the digital signal, the current is decoded and interpreted : artificial biotopes appear and grow on the screen. "A-Volve" (1993-1994) is an interactive work displaying virtual creatures in a glass container filled with water. Forms and movements are connected : influenced by the environment, the creatures change shape in real time. They are generated by spectator, who draws shapes with an electronic pen on the screen, there by creating three-dimensional organisms which are immediately transmitted and represented as living creature in the water. In "Phototropy" (1994) the virtual organisms whose existence depends solely on the light are generated by a computer.

In the piece "The artificial life Metropolis CELL" (1994) Yoichiro Kawaguchi presents three-dimensional plastic art as represented by a computer image using the basic principle of a cellular model as seen from an artistic point of view. The self-organisation of a three-dimensional space, which is composed of a mass of "voxel", can generate complex evolving space. This idea is inspired by a the theory of self-organising automaton in two-dimensional space. A model based on a three-dimensional cell model is capable of inducing unpredictable, delicate and emergent vibration using a simple rule. A group of "voxel" makes up a three-dimensional object, just as millions of cells make up a natural life form.

Artists create personages whose evolution can be modified by the spectator.
"Kaleidoscope" by Jaffrenou, is a fictitious world, inhabited by mythical personages and legendary, whose evolution can be modified by the spectator. Jaffrenou is thinking of creating virtual worlds, outside the video screen frame, reflecting real world events, or, at the contrary, purely imaginary ones, eventually created by electrons. Driving radio-controlled vehicles, the spectators would discover another Universe.

Self-identification is a important trend of Cyberart.
"Places holder"(1994) was realized by Brenda Laurel, a researcher working on the relation between culture and new technologies, in collaboration with Rachel Strickland artist. "Places holder" explores a new paradigm for narrative action in virtual environments. It allows the spectators to feel the same impression as their ancient ancestors. Using head mounted display, the spectator moves in a visual and sonorous environment inhabit by Indian tribes ritual animals – snakes, spiders, fishes and crows. The experience took place in a recreated prehistoric forest. The participants had not only to see and hear, but also to move as the animals they were representing. Doing so emphasized the identification.

10.3. Conclusion

Although Cyberart is, obviously, the most representative form of art, in this day and age. Its development belongs to the future. The artist working on Cyberart, have in common their will to explore, in their work, a wide range of aesthetic categories, using advanced technologies. Meanwhile, they can be distinguished from the previous artist generations, using equally technologic process, and in sometime, they can be distinguished from their own previous works. Indeed, they have realized the enormous sociocultural changes issued from the technologic progress. During the eighties and the nineties, these artists have endeavored to establish a significant connection between the fundamental human experiences - both physical, psychological and intellectual - and the radical global intrusion of cyberspace and cyberculture, in every life domains, with its accompanying beneficial effects, potential dangers and incredible possibilities.

The state of Cyberart today is comparable to the state of cinema at the beginning of our century. Moreover, the cinema, went beyond its original technologies and summoned up its world and our dreams. Above all, the cinema was and is still, a change from the piece to the imaginary. As Andre Bazin outlined it, the cinema is the materialization of a made and displayed image, in and through the space time. It is a certain mental vision – the idea of a "total realism"- the inventors are going to offer it the keyboard and the visual sonorous games of some technical systems. The authors are going to play with it.

Today, the artists have just started to explore the expression means of Cyberart. They are previewing the direction of its development. Its language, its traditions, and its genre still have to be invented.

References

Boissier, J.L., 1991, Machines à communiquer faites œuvres, *Qu'est-ce que la communication* (Paris).

CNAM, http://www.cnam.fr.

Costa, M., 1991, Technology, Artistic production and the Aesthetics of communication, *Leonardo*, **24**.

Couchot, E., 1990, Je sème à tous vents, *Artifices: invention, simulation* (Saint Denis), 38.

Couchot, E., 1992, Une marge étroite mais fertile…, *Virtual Review*, **1**.

DeKerckhove, D., 1991, Communication Arts to Have New Spatial Sensibility, *Leonardo*, **24**.

Forest, F., 1993, Pour qui sonne le glas ou les impostures de l'art contemporain, *Quaderni*, 123.

Galloway, K., http://www.net.com, http://www.tmn.com/0h:cmmunity/kitchen, http://www.uni-c.dk/80/ecafe.

Housz, I., 1995, Interview of Woody Vasulka, *Technikart*, **16**, 44.

Kac, E., 1992, Sur la notion d'art en tant que dialogue visuel, *Art-réseaux*, Editions du CERAP (Paris), 23.

Knowbotic, http://www.khm.uni-koeln.of/kr+cf/knowbotic-south.html.

Legrady, G., 1990, Image, Language and Belief in Synthesis, *Art Journal* (New York), 266.

Musso, P., 1993, Interview of Piotr Kowalski, *Quaderni*, **21**, 99.

Popper, F., 1993, *L'art à l'age électronique*, Hazan (Paris).

Stelarc, http://www.merlin.com.au/stelarc.

Utopia, http://www.igc.apc.org/femacart.

Waxweb, http://bug.village.virginia.edu.

Authors Index

Index

C

J

K

L

M

N

O

P